John Traherne Moggridge

Contributions to the flora of Mentone and to a winter flora of the

Riviera

Including the coast from Marseilles to Genoa

John Traherne Moggridge

Contributions to the flora of Mentone and to a winter flora of the Riviera
Including the coast from Marseilles to Genoa

ISBN/EAN: 9783337257392

Printed in Europe, USA, Canada, Australia, Japan

Cover: Foto ©berggeist007 / pixelio.de

More available books at **www.hansebooks.com**

CONTRIBUTIONS

TO THE

FLORA OF MENTONE

AND TO A

Winter Flora of the Ribiera,

INCLUDING THE COAST FROM

MARSEILLES TO GENOA.

BY

J. TRAHERNE MOGGRIDGE, F.L.S.

LONDON:
L. REEVE & CO., 5, HENRIETTA STREET, COVENT GARDEN.
1871.

LONDON:
SAVILL, EDWARDS AND CO., PRINTERS, CHANDOS STREET
COVENT GARDEN.

PREFACE.

THE want of an illustrated Continental FLORA has long been felt by tourists, invalids, and others, who fail, either from want of power or inclination, to determine their plants by the present available means. Though unable at present to commence such an undertaking, I hope that the present work may afford some facilities which may induce not a few invalids and others to turn their attention to the study of the wild flowers of the district, and thus find a pleasant subject for recreation. When considering the thousands of idle hands which every winter pull myriads of flowers to pieces south of the Alps, and the thousand restless energies all craving for employment in weary satiety of absolute rest, it becomes quite a marvel that these hundred-handed colonies of English should so rarely be set to work at drawing for publication some few of the wonderful objects of Natural History by which they are everywhere surrounded. In the water, the earth, the air, unknown wonders await diligent search and investigation, while the host of things half-known teems with opportunity for scientific inquiry. Well-directed research in any definite direction must afford happy employment for the invalid, and tend towards the advancement of knowledge. Being therefore convinced that I had better

do a little as well as I could, than sit down and lament that that
little was indeed so small, I began collecting the drawings for the
present work, in the hope that I might lay a pebble towards founding
the great illustrated FLORA OF EUROPE, which I long to see
commenced in earnest, and with the intention of offering to the
lover of Nature some account of a few of the rarities and beauties
by which I was every day encompassed. I had great advantages
before me, such indeed as can rarely, I think, present themselves
to many. In the first place, my father was indefatigable in pro-
curing subjects for my pencil, his knowledge of plants and great
powers of endurance making him as able a collector as ever searched
jungle or climbed Alp. Besides his all-important help, I was most
deeply indebted to M. Honoré Ardoino, who, though himself engaged
upon a Flora of the Department, spared me both time, and specimens
from his valuable herbarium, at once becoming one of the readiest
promoters of my scheme. I take the present opportunity of thanking
him most warmly for his kindness and liberality. With the aid of a
catalogue of the plants of the neighbourhood, published by him in
1862, it was easy for me to work up the descriptions in the "Flore
de France," of Grenier and Godron, and a few other books, of the
greater number of Mentonese plants. I hope it will not be long
before M. Ardoino will give to the world his "Flore du Département
des Alpes Maritimes," which will include Cannes, Nice, and Mentone. The student of Botany will then have no difficulty in naming
most of the plants likely to be discovered in this last corner of South
Eastern France.

The descriptions of the plants figured in the present work are, of
course, for the most part compiled from a variety of authors, every
part being, however, checked by comparison with the actual plant.
I have sometimes been obliged to make more guarded statements and
occasionally to differ altogether from my authorities. In all cases my

drawings were made from fr eshly-gathered specimens, so that the peculiarities of the lines which give so much character to different styles of growth might be if possible rendered. All the dissections are represented as being magnified to a greater or less degree, except in a very few instances when special mention is made to the contrary in the accompanying letter-press.

I can only hope that the Reader may find that an ever-increasing, ever-widening interest attends his researches, till many other branches of science link themselves into his original study, and make the pursuit endless though never in vain.

September, 1864.

NATURAL ORDER RANUNCULACEÆ.

Tribe—ANEMONEÆ. Section of genus Anemone, having carpels without tails, and bracts of the involucre sessile.

PLATE I.—*Anemone pavonina. De Candolle, inclusive of β— Woods, inclusive of β— Anemone hortensis β. γ. of Grenier and Godron.*

GENERIC.—*Petals* o. Calyx petaloid, generally of from 5 to 15 sepals. *Involucre* three-leaved, distant from flower.

SPECIFIC.—*Carpels* woolly. *Sepals* scarlet, either 10 or more in the single form, but as the flower becomes more double, the sepals take a narrower form, till when the stamens are obliterated, they are linear lanceolate. *Involucre* sessile, of three lobes, either notched or entire. *Leaves* radical, divided into three wedge-shaped lobes, more or less cut at the edges.

EXPLANATION OF PLATE.—Plate I. represents the extreme forms under which this plant is found at Mentone, and it is well to remark that every possible stage may be observed between the two.

REMARKS.—I believe that some most competent observers have shown that this plant is not distinct from the lilac variety figured at Plate II., and I am told that the variety Anemone versicolor (Jordan), which is found at Grasse, completely re-unites them. It is rather remarkable that here, and at Nice, where A. pavonina is of more variable form than elsewhere, the colour remains, as far as I can learn, quite true and unchanged. There is another Scarlet Anemone to be found here, but that may be at once distinguished by its leaves, which are so finely cut as to have earned it the name of the Parsley-leaved A. ; it moreover lacks the fine yellow ring which surrounds the stamens of the present species. The specimens from which I have drawn were gathered in the Turin Valley in the early part of March. Their time of flowering is from the beginning of February to the end of March. I have never seen this plant growing at any distance from cultivated ground, where it soon becomes a well-established weed.

NATURAL ORDER RANUNCULACEÆ.

Tribe—ANEMONEÆ. Section of genus Anemone, having carpels without tails, and bracts of the involucre sessile.

PLATE II.—*Anemone Stellata Lamarck.* *A. hortensis Woods.*
A. hortensis var. a. of Grenier and Godron.

GENERIC.—*Petals* o. *Calyx* petaloid, generally of from 5 to 15 sepals. *Involucre* three-leaved, distant from flower.

SPECIFIC.—*Carpels* woolly. *Sepals* definite in number ; either linear-acute or lanceolate obtuse in form ; varying in colour from bright lilac to pure white. When fairly expanded, the flower has a diameter about equal that of a half-crown. *Involucre* of three sessile bracts, generally but little divided. *Leaves* small, and often but little divided, forming three wedge-shaped lobes.

EXPLANATION OF PLATE II.—This represents the common form of A. stellata, near Mentone, with leaves of variable form, but approaching those of A. pavonina.

REMARKS.—I have never seen any Anemone which could be confused with this, and I believe that, if Mentone was its only habitat, it might certainly be reckoned a good species. Unfortunately, between 20 and 30 miles off (at Grasse), there exists a hot-bed of varieties, which, as we have seen, (page 1, Remarks,) prevents our so considering it. The specimens figured came from the olive grounds near Cap Martin, where the plant abounds throughout March and the early part of April. This Anemone is not exclusively found in cultivated places, and may be seen in a very wild and rocky situation underneath the ramparts of Monaco.

2

NATURAL ORDER RANUNCULACEÆ.

Tribe—PEONIEÆ.

PLATE III.—*Pæonia peregrina.* *De Candolle syst. & fl. fr.*
Grenier and Godron. *Woods.*

GENERIC.—*Flowers* regular. *Follicles* 2 to 5, many-seeded, bursting inwards. Petals 5 or more.

Sepals 5, persistant, unequal.

SPECIFIC.—*Carpels* covered with whitish down, upright and separated when ripe.

Anthers not half the length of the filament. *Petals* 5 to 10, obovate, obtuse. *Leaves* biternate and ternate, the middle lobe of 3 or 5 decurrent segments. The backs of the leaves whitish grey, with downy pubescence.

EXPLANATION OF PLATE.—This drawing of Pœonia peregrina must be taken as representing a small flower, and, of course, but a few of the uppermost leaves. The diameter of many blossoms was from 5 to 6 inches. Fig. 1 represents an immature fruit. Fig. 2 is of a stamen; this fig. and fig. 1 are of the natural size.

REMARKS.—As yet no other species of Peony has been discovered at Mentone; and if there were, I think that the whitish backs of the leaves would sufficiently distinguish this. Though very rare as an European plant, it abounds on the summits of the more lofty mountains, at a height of from 3,250 to 4,000 feet. The localities cited by Grenier and Godron, in their "Flore de France," as habitats are as follows :—Serane, at the foot of the St. Loup peak, the wood of Valène, near Montpellier; a wood at Die; Cévennes; Mende; Grasse; Rousillon, at Abeillas, near Bagnols-sur-Mer; Perpignan. It would be very interesting to know the respective heights of the places enumerated, but I have no means of gaining the information. We see a good example in the present species of the fact that it is not in the garden only that plants are able to produce strange and variable forms; the number and shape of the petals being so irregular, that it is rare to find any two flowers alike. The specimens figured were gathered for me on the Mulaciers mountain, where they were blossoming in great profusion on the 7th of May.

Tribe—ALYSSINEÆ. Section of genus Alyssum, having simple filaments.

PLATE IV.—*Alyssum halimifolium. Linn.—Koniga halimifolia. Woods—Lunaria halimifolia. Allionii.*

GENERIC.—*Pod* opening at a partition on its broadest diameter (Latisept), containing a few pendulous seeds with accumbent cotyledons. *Stamens* frequently with glands or notches at the base of their filaments. *Petals* equal.

SPECIFIC.—*Style* much shorter than the ripe pod. *Seeds* provided with a wide ring of membranous texture, placed in orbicular pods which form a compact cluster when ripe. *Stamens* of equal height, having anthers of an oval shape. *Petals* broad in the limb, then abruptly narrowed into a short claw.

EXPLANATION OF PLATE.—Plate IV. represents Alyssum halimifolium with flower and fruit; the latter being extremely difficult to procure ripe, on account of the rapidity of ripening. Fig. 1, the seed with its wide membranous wing. Fig. 2, the accumbent cotyledons, showing how the two seed-leaves lie parallel to one another under the radicle. Fig. 3, a flower from which the petals are taken away, displaying the equal stamens, and the calyx and peduncle sparingly covered with stellate hairs. Fig. 4, a petal showing the sudden contraction of the limb. Fig. 5, part of a leaf densely covered with stellate hairs, which give to the unassisted eye an impression of greyness.

REMARKS.—With regard to the description of this genus, I have ventured to omit the statement, made by most authors, that the valves of the pods are not veined, as I found both nerves and veins in those of the specimen here represented. Although this species is rare generally, it is sufficiently common on the higher mountains about Mentone; and while never itself braving the heat of the shore, is there represented by the common sweet-scented little Alyssum Maritimum. From this latter plant the present is easily distinguished by its broad pod; while the wide membranous wing round the seed separates it from the more nearly related species A. perusianum and A. spinosum. The Honesty of our Gardens is a member of the Alyssum tribe, and Allionii considered this genus (Lunaria) should include A. halimifolium. The specimen figured came from the Berceau mountain, the 9th of November, on which mountain and on a few others, plants may be found at an elevation of not less than 2,150 feet. Late blossoms may be gathered till the middle of November in late seasons, and in the end of April flowering recommences.

NATURAL ORDER LINE.Æ.

Section of genus Linum, having free petals, and leaves without glands at their bases.

PLATE V.—*Linum Maritimum. Linn. Woods. De Candolle. Grenier and Godron.*

GENERIC.—*Styles* 5, rarely 3, terminating in an ovary, whose cells equal or double their number. *Stamens* 5, rarely fewer, sometimes 10. *Petals* equaling the stamens in number. *Sepals* equaling the stamens in number.

SPECIFIC.—*Stigmas* brush-shaped. *Styles* generally twice as long as the stamens, but in individual instances shorter than them? *Petals* of a fine sulphur yellow, about three times as long as the calyx? *Sepals* broadly obovate, acuminate, fringed with glands. *Peduncles* arranged in somewhat scorpioid racemes.

EXPLANATION OF PLATE V.—This plate represents Linum maritimum, one of the many beautiful representatives of the family to be found in the neighbourhood. Fig. 1 shows the calyx, with the stamens in the act of emitting their pollen, and the five brush-headed styles pushed out high above them.

REMARKS.—This Flax genus is destined from henceforward to bring to our remembrance the name of Mr. Darwin. His wonderful discoveries concerning the reciprocal fertilization in certain species of the genus Linum, not only are offered as startling facts to men of deep and established knowledge, but are presented to you and to me, my Reader, that we may have fresh subjects for inquiry, and definite objects on which we may expend our powers of observation. Mr. Darwin has proved (for account refer to "Journal of the Proceedings of the Linnean Society" for May 13th, 1863, page 69, and following pages) that in some species of Linum—as, for example, Linum grandiflorum—the pollen of the flowers of any one given plant is impotent to fertilize those flowers, and that any such plant, without receiving pollen other than its own, would set no capsules productive of good seed. But Linum grandiflorum is dimorphic; that is to say, produces two kinds of flowers, each on a separate plant, one with long styles, as in fig. 1, and the other with styles so short as to be hidden by the stamens. Now, the pollen from

PLATE V.

the long-styled blossoms will fertilize completely the stigmas of a short-styled plant, which then bears plenty of good seed, and *vice versâ*. This is what I meant by reciprocal fertilization. He tells us also that some of the commoner species have not their styles of different length, but that, when plants are so formed, the two kinds, each of which is so essential to the welfare of the other, are to be raised from seed in nearly equal quantities. I regret that I was unable to re-examine Linum maritimum with this great fact before me, especially as Mr. Darwin in that same paper quotes Planchon ("Hist. Physiolog. des Plantes d'Europe." 1841. Tom. i., p. 401) to the effect that this very plant, as well as L. Gallicum and Strictum, both of which are found at Mentone, is dimorphic. I can only state, as some guide for further and fuller investigation, that I have found but one specimen in which the styles were shorter than the stamens, though great numbers of specimens have passed through my hands. I notice also that plenty of capsules are formed and ripened.

My figure was taken from plants brought from Ventimiglia in November, where it continues to blow until the first severe cold, and re-appears again about March. Grenier and Godron give Corsica, the Mediterranean coast, and, northwards, to Avignon, as its district; it was, moreover, found on the shores of Algiers and Tunis. (Desfontaines' " Flora Atlantica.")

NATURAL ORDER MALVACEÆ.

Tribe—MALVEÆ. Section of genus Lavatera, having but one peduncle in the axil of a leaf.

PLATE VI.—*Lavatera Maritima. Gouan. De Candolle. Woods. Grenier and Godron.*

GENERIC.—*Stigmas* filiform. *Inner calyx,* five-lobed. *Outer calyx* of one leaf, deeply three-lobed.

SPECIFIC.—*Carpels* large, blackish when mature, having waved edges prolonged into wings, and flat backs. *Axis* prolonged into a smooth conical apiculum, not spreading over the carpels. *Petals* obovate, twice the calyx in length, of a pale lilac, with a dark-coloured, hairy fringed claw. *Inner calyx* lobed triangularly, covering the carpels after the withering of the flower. *Outer calyx* lobed lanceolately, the lobes being sometimes reflexed. *Peduncles* generally longer than the leaves. *Leaves* densely covered with stellate hairs, giving the plant a greyish appearance and velvety texture. These hairs are constantly wearing away, and cling together in tiny groups on the stems and edges of leaves, in the manner of wool. *Growth* that of a shrub, with a low woody stem, two feet high.

EXPLANATION OF PLATE VI.—Plate VI. shows Lavatera maritima, as it may be gathered about Christmas time at Mentone. Figs 1 and 2 represent the seed, fig. 2 of the natural and fig. 1 of the magnified size. The wrinkled surfaces and winged edges should be noticed. Fig. 3 is of the claw of a petal, with its hairy fringes.

REMARKS.—Lavatera maritima grows along the French and Italian shores of the Mediterranean ; in Sardinia ; in Spain (Cavanille's " Diss. Bot.") ; and in Algiers (Desfontaines' " Fl. At."). At Mentone I have rarely seen this plant growing at any distance from the Jurassic lime-stone, where it delights to wedge itself into the cracks of the hardest and most inhospitable rocks, or to vie with Euphorbia dendroides in endurance of the heat of arid and scorching situations. The natural order Malvaceæ includes many well-known and highly-valued genera, such as Cotton (Gossypium), Hollyhock (Althea), Hibiscus, Abutilon, Sida, &c. In the year 1863, a small plot of Cotton was grown experimentally at Mentone ; and I had the great pleasure of seeing plants

PLATE VI.

in every stage, from the opening of the flower, to the shedding of fleecy masses of Cotton-wool from the gaping capsules. I regret that I have no means of judging whether the experiment was a success or no, and that I am, moreover, unable to name the species of Cotton grown. The flowers were about four times the size of those of Lavatera, and of a pale sulphur yellow; the corolla was never widely expanded, and the stigma was entire, with dotted channels on its surface. The whole plant was about the height of a raspberry bush, but with a woody branching stem, and the upper parts rough with coarse hairs. I believe that Gossypium barbadense and its varieties yields the best Cotton of manufacture. The specimens from which I drew this plate were gathered in January, the seeds being procured rather later. Flowering commences about Christmas time, and ends in April.

3.

2.

NATURAL ORDER TEREBINTHACE.E.

PLATE VII.—*Cneorum tricoccon. Linn. Woods. Grenier and Godron.*

GENERIC.—*Fruit* composed of from 2 to 4 nuts, each of which encloses 2 ovules, separated by an imperfect partition. *Stamens* 3 or 4. *Petals* 3 or 4, placed below the hypogynous disk. *Flowers* hermaphrodite.

SPECIFIC.—*Style* persistant. *Petals* oblong, bright yellow. *Calyx* minute, with obtuse lobes. *Flowers* on short axillary peduncles, bearing two small bracteoli. The peduncles sometimes are clustered in the axil, and always spring from the uppermost leaves. *Leaves* alternate, smooth, oblong, quite entire. *Growth* that of a small evergreen shrub, from 2 to 3 feet high, having a strong and disagreeable smell.

EXPLANATION OF PLATE VII.—The preceding page describes Lavatera Maritima as a winter-flowering plant, and Cneorum tricoccon is no less one. It should be noticed that blossoms having their parts in fours are represented on the same branch with those in threes. This is actually the case, and may also be observed in the fruit. Fig 1 is of a flower with its bracteoles attached against the lower part of the peduncle. Fig. 2, the same, deprived of its petals, showing the hypogynous disk. I remark that at a later stage the stigma separates into three lobes. Fig. 3, the fruit cut longitudinally, showing the two cells empty.

REMARKS.—Cneorum tricoccon is found in dry parts of the French coast of the Mediterranean, and is cited as growing in Tunis and Algiers by Desfontaines. The localities it chooses at Mentone are much the same as those assigned to Lavatera Maritima. The Natural Order Terebinthaceæ, in which this plant is placed, includes many plants from which we derive valuable resinous juices, and, in a few cases, edible fruits. At Mentone we have, as representatives of the order, in addition to Cneorum, the common Sumach (Rhus Cotinus), with Pistachia lentiscus and terebinthus. These two last-named trees yield respectively mastic varnish and chio-turpentine. The mango (Mangifera indica), the hog-plum of the West Indies (Spondias purpurea and S. Mombia), and the Cashew nut (Anacardium occidentale), are among the edible fruits mentioned above. I remember questioning an Indian officer about the Mango, and his replying that the best quality was very hard to procure, and that inferior ones were stringy and *tasted of turpentine;* so the latter deserves a place in an order of Turpentine plants. My figure was taken from specimens gathered in January, but more or less blossom may be found throughout the winter and spring.

PLATE VIII.—*Lupinus reticulatus. Deso. Grenier and Godron.
Not given by Woods.*

GENERIC.—*Style* subulate, ascending, having the stigma slightly bent
forward. *Pod* large, oblong, divided in the interior by partitions forming shallow cavities containing the seeds. *Seeds* having the funicle
narrow at the hilum. *Stamens* monadelphous. *Calyx* deeply divided
into two separate lips. *Leaves* digitate, having stipules united to the
petiole.

SPECIFIC.—*Pod* wavy on the upper edge. *Seeds* 5 to 7, speckled with
grey, and marked with black and white streaks. *Calyx* furnished on
either side with a small linear appendage ; upper-lip short, bifid.
Flowers alternate, bright purplish blue. *Leaves* linear-obtuse, channelled.

EXPLANATION OF PLATE VIII.—This plate represents Lupinus reticulatus, a plant about which there has formerly been some confusion. Fig.
1, half of a nearly ripe pod of the natural size, showing the partitions,
with seeds in them. Fig. 2, Calyx, with the ovary removed, showing
the linear appendage. Fig. 3, the lanceolate, deciduous bract, which
falls as soon as the flower begins to open.

REMARKS.—There is no other Lupine known as growing at Mentone,
but I believe that the present species may be distinguished from any
other by its narrow channelled leaflets. Papilionaceous plants are generally considered as a sub-order of that great Leguminous Order, which is,
after the Synantheraceæ or Composite plants, the largest in the world.
Among the productions of the present sub-order are the indigo of com-
merce (Indigofera tinctorea and cærulea), liquorice (Glycyrrhiza glabra),
peas, beans, clover, &c. ; also, many showy garden plants, as Laburnum,
Wistaria, Robinia, &c. The two other sub-orders are the Cæsalpineæ and
Mimosæ. The former contains the Cassia, which yields the senna of
the drug-shops, and the Carob-tree (Ceratonia siliqua) to be seen commonly at Mentone ; the latter, Acacias, species of which (A. vera and
arabica) yield gum-arabic, and the genus Mimosa. from which the
sub-order is named. The specimen figured was brought to me on April
7th, when the flowering was commencing ; the pods were gathered in
May.

NATURAL ORDER LEGUMINOSÆ.

Tribe—PAPILIONACEÆ.

PLATE IX.—*Coronilla valentina Linn. C. Stipularis. Lamarck. De Candolle. Woods.*

GENERIC.—*Pod* jointed, cylindrical, nearly straight. *Stamens* having the longer filaments dilated at the summit. *Keel* beaked, acuminate. *Calyx* short, campanulate, having the uppermost teeth partly united. *Leaves* imparipinnate, rarely trifoliate.

SPECIFIC.—*Pods* drooping, nearly straight, of about seven, more or less warted joints. *Standard* having an obovate recurved limb, and a short claw without a scale. *Peduncles* often longer than the leaves, bearing pedicels longer than the Calyx. *Stipules* orbicular, mucronate, quickly falling. *Growth*, that of a much-branched, low-growing shrub, with a strong woody stem ; the whole plant averaging about 3 feet.

EXPLANATION OF PLATE IX.—This drawing represents Coronilla valentina with its curious jointed pods, and a piece of one of the woody branches cut tolerably high up.

REMARKS.—This Coronilla can scarcely be confused with any other plant growing at Mentone. Coronilla Emerus is the only other representative growing there as a shrub, and that species has pendant flowers, a scale on the interior of the standard, and grows in moist shady places, instead of under the scorching limestone rocks chosen by the present species. In the gardens, and in some stations in France, is found a plant very nearly resembling this, named by Linnæus Coronilla Glauca. It may be distinguished, I believe, by its having small linear stipules, in place of the great orbicular ones found in C. valentina. When Grenier and Godron wrote their "Flora," no habitat was known for this rare and beautiful shrub in France, and their only locality was at St. Florent, in Corsica. Near Mentone this plant grows in three or four spots, and is most plentiful in the neighbourhood of the Pont St. Louis, where the specimens figured were gathered March 9th, 1864. The jurassic limestone is the only formation on which I know of it, and its favourite soil is formed by the *débris* from some southern facing cliff, among the fissures of which it may strike its roots.

NATURAL ORDER ROSACEÆ.

Tribe—DRYADEÆ. Section of genus Potentilla, having woody stems.

PLATE X.—*Potentilla saxifraga. Ardoino in his Catalogus des Plantes Vasculaires de Menton. Lehman. Not given by Woods.*

GENERIC.—*Style* lateral or nearly terminal. *Fruit* composed of many nuts on a flattish, dry receptacle. *Seed* pendulous or ascending. *Stamens* numerous. *Petals* 4 to 5. *Calyx* concave, 10-parted, in two series, five exterior parts smaller. Babington's Manual.

SPECIFIC.—*Nuts* concealed in their silky hairs. *Receptacle* densely hairy. *Stamens* smooth. *Petals* white, subrotund, longer than calyx. *Segments of Calyx* acuminate, ciliate with glandular hairs. *Peduncle* bearing many flowers, in a corymbose inflorescence, which springs from the axils of trifoliate floral leaves. *Leaves* ternate or quinate ; the leaflets being elliptic-lanceolate toothed or entire, coriaceous, glabrous above and silky beneath, margins revolute, not ciliate. *Stipules* acuminate, united to petiole throughout the greater part of their length. *Stem* very short, woody ; its branches concealed in the dead matted stipules, and beneath the dense covering of the imbricate living ones. *Growth* woody, forming dense tufts, resembling those of Dryas octopetala ; and though not herbaceous, has not the habit of a shrub.

EXPLANATION OF PLATE X.—This plate represents Potentilla Saxifraga, and is, I believe, the first coloured drawing of this rare and lately discovered plant that has been published. Fig. 1, the immature nut, taken from the flower with the style. Fig. 2, the flower with the petals removed. Fig. 3, a stamen showing the anther previous to the emission of pollen.

REMARKS.—Potentilla Saxifraga was discovered by M. Ardoino, and in his "Catalogue" we find the following account of its habitats :—"I found this beautiful species in 1847, upon the precipitous rocks between St. Agnes and Castillon, at 870 mètres (equal to about 3,480 feet) ; it has been since re-discovered at Peglia, Duranus, Raus, and other points of the Alpes Maritimes." The plant has, within the last year or two, been discovered on other mountains near Mentone, and in the celebrated Gorge de Saorge, on the road from Nice to Turin. On heights near the shore it chooses northern exposures, not tolerating the heat of the direct

PLATE X.

sun. Its woody stem distinguishes it at once from the other species found at Mentone ; and, as far as I know, Potentilla fruticosa, with its tall branching growth and yellow flowers, is the only European species sharing with it this peculiarity. The specimen figured was procured for me in the Gorge de Saorge, April 23rd, when a few stray blossoms were coming out. I believe that this Potentilla does not come into full blow till the middle or end of May, but occasionally small blossoms may be procured in the spring and autumn.

NATURAL ORDER PRIMULACEÆ.

Tribe—PRIMULEÆ. Section of genus Primula, having a calyx much shorter than the tube of the corolla, and the young leaves rolled inwards.

PLATE XI.—*Primula marginata. Curtis. Woods. P. crenata Lamarck. De Candolle.*

GENERIC.—*Capsule* many celled, with five entire or bifid teeth.

Stamens 5, opposite the lobes of the corolla, on the tube of which they are inserted.

Corolla salver-shaped, with a cylindrical tube. *Calyx* campanulate, or tubular.

Flowers dimorphic?

SPECIFIC.—*Flowers* bright lilac, mealy or not at the throat, dimorphic.

Involucre composed of oval phyllaries. Capsule sub-globular, about equal the calyx. *Calyx* with rounded teeth, the margins of which are covered with white meal. *Leaves* oboval, having their crenately-toothed margins edged with white meal. *Plant* free from viscidity, or any kind of pubescence; abounding in mealy powder.

EXPLANATION OF PLATE XI.—Plate XI. represents Primula marginata, a plant named by Curtis in the "Botanical Register," where a figure is given (vol. lxvii., p. 191). Figs. 1 and 2 show the calyx and part of the tube of the corolla taken from two distinct plants, and exemplify the two forms or dimorphic condition. Fig. 1 is of the long-styled form; fig. 2. of the short-styled.

REMARKS.—I must again attempt to give a very brief outline of some of the facts recorded by Mr. Darwin, (see "Journal of Proc. of Linn. Soc.," March 8th, 1862,) with reference to this genus; not that any such mere sketch can suffice to give any adequate conception of his many careful observations, so laboriously made and so concisely reported; but that perhaps this notice may afford some guide to those who have not studied the paper, and might otherwise bestow no second glance upon Pin or Thrum-eyed primroses.

Pin and Thrum-eyed primroses are familiar to most children, where in the former the stigma, and the latter the stamens, are seen in the throat of the corolla. Now, it seems very possible that the whole genus Primula is in the same condition, and Mr. Darwin, in the paper before

PLATE XI.

alluded to, shows us, by the following comparison, how Pin-eyed plants bear less seed than the Thrum-eyed :—"The long-styled plants have a much longer pistil, with a globular and rougher stigma standing high above the anthers. Stamens short, grains of pollen smaller, oblong in shape ; upper half of the tube of the corolla more expanded ; number of seeds produced is smaller." "The short-styled plants have a short pistil, half the length of the tube of the corolla, with smooth depressed stigma beneath the anthers. Stamens long, grains of pollen spherical and larger. Tube of the corolla of the same diameter till close to its upper end." In summing up his observations he tells us that in the case of Primula veris, auricula and sinensis, all of which are dimorphic, the pollen of the opposite form in each of these plants is most essential for their perfect fertilization ; and that this reciprocal benefit conferred, doubtless favours the intercrossing of species at the same time. He makes mention of some dimorphic Primulas, and casually refers to P. Marginata as probably in that condition. I have to regret that I had no knowledge of the above facts when I made my drawings, and the dissections which I have given, though generally trustworthy, are not as detailed as I could wish. In Part II. of this work, I have a third Primula to figure (P. Allionii), when I shall hope to add some notice of such facts as I have been able to gather on all three species. P. Marginata grows near Mentone on the higher mountains, at an elevation of from 3,400 to 4,000 feet. The specimens figured were gathered April 15th.

Tab. II.

Tribe—PRIMULEÆ. Section of genus Primula, having a calyx much shorter than the tube of the corolla, and the young leaves rolled inwards.

PLATE XII.—*Primula latifolia. Lapeyrouse. Woods. Grenier and Godron. P. viscosa. Allionii.*

GENERIC.—See description of preceding plate.

SPECIFIC.—*Flowers* purple, dimorphic, from 5 to 20 in number. *Capsule* globular, longer than calyx. *Leaves* oboval-oblong, slightly crenate above, gradually narrowed into the petiole ; densely covered with glandular hairs. *Growth* that of an herbaceous perennial, with large tufted fleshy leaves, which are sticky to the touch, and bearing a stout peduncle generally longer than the leaves, making the plant from three to six inches high.

EXPLANATION OF PLATE XII.—Plate XII. shows a short-styled plant of Primula latifolia. Figs. 1 and 2 exemplify the two forms in this plant ; fig. 1 being part of a blossom taken from the short-styled plant drawn in the plate, and fig. 2 the long-styled form in a distinct specimen of Primula latifolia. In my one specimen of this latter form, the stamens were not, as is usual in such other Primulas as I have seen, sunk in the bottom of the tube, but stood at about a quarter of its height.

REMARKS.—By referring to the preceding plate and its description, some short account will be found of Mr. Darwin's discoveries with regard to some species of this genus, as recounted in the "Linnean Journal." This beautiful Primula has not yet been found in the immediate neighbourhood of Mentone, and the specimen from which I have drawn was one of several brought me from near Sospello, in the valley through which the road runs from Nice to Turin. It cannot be confused with any other species that I have yet procured at Mentone, but greatly resembles Primula villosa of De Candolle, which is chiefly to be distinguished by its obovate or nearly orbicular leaves, which are suddenly narrowed into their petioles. In a monograph of the Primulas by Lehman, I find the following habitats given : the Alps of Carinthia,

PLATE XII.

Styria, the Tyrol, Piedmont, Dauphiné, and the Pyrenees. It is worthy of notice that the Primrose (P. vulgaris) is the only one of the tribe that endures the heat of the shore level; while it is rare that the cowslip (P. veris), or the Oxlip (P. elatior), which some do not consider specifically distinct, descend below 2,800 feet, a height rarely gained by the primrose. Primula latifolia was in fine blossom on March 29th in the year 1864, and had, by its appearance, been sometime in flower.

Tribe—CONVOLVULEÆ. Section of genus Convolvulus, in which the plants are annual, and do not climb.

PLATE XIII.—*Convolvulus siculus.* *Linn.* *Woods.* *Grenier and Godron.*

GENERIC.—*Stigmas* two on a simple style. *Capsule* two or four-celled, each cell containing one or two seeds. *Corolla* campanulate, having five angles and five folds.

SPECIFIC.—*Capsule* smooth, on recurved peduncles. *Corolla* bright purplish blue, about twice the length of the calyx. *Bracts* placed close to the calyx. *Calyx* hairy, divided into pointed segments of variable shape. *Stem* ascending, downy, not climbing on other plants. *Leaves* pubescent, somewhat heart-shaped, with more or less prominent auricles. *Growth* that of an annual, with a feeble stem.

EXPLANATION OF PLATE XIII.—This drawing of Convolvulus siculus is given, not only on account of its beauty and rarity, but also with a hope that some other Mentonese habitat may be found, beside the one where alone it has yet been discovered.

REMARKS.—The specimens from which this plate was drawn were gathered on April the 21st, in the celebrated Gorge of St. Louis, below the Corniche road, where, among the crevices of the precipitous limestone rocks, these lovely plants are sparingly scattered. Convolvulus siculus is one of the rarest plants of this latitude, being in truth, as its name imports, a Sicilian and a Southerner. There is no other species known to me with which the present can be confused, Convolvulus tricolor, so common in gardens, being one of the nearest to it. Many other plants of this order are well known for their beauty, and have long been cultivated; among these the Ipomæas, Pharbitis, and Quamoclit, stand pre-eminent. Some also, as Ipomæa purga, and Convolvulus scammonium, yield drugs; the former, which grows at Xalapa, in Mexico, affording jalap, and the latter, a native of Syria and the Levant gives a root from which Scammony is extracted. Convolvulus siculus is given by Grenier and Godron as growing at Toulon and in Corsica, and

PLATE XIII.

is also found at Nice and Monaco, finding on this French shore of the Mediterranean its northern limit.   It is cited by Desfontaines as seen in the sands of Algeria, where a white variety is also mentioned.

The time of flowering is from the end of April to the middle of May, appearing about the same time as that splendid plant Convolvulus althœoides, whose large blossoms attract universal attention.

2   3

1

NATURAL ORDER VERBENACEÆ.

PLATE XIV.—*Vitex Agnus-castus. Linn. Woods. Grenier and Godron.*

GENERIC.—*Fruit* a kind of drupe, apparently formed by the enclosure of four nuts (or fewer by abortion) in a dry corky mass, which is encased in a hard globular shell. *Stamens* four, exsert, didynamous. *Corolla* composed of a short tube, with a five-lobed, bilabiate limb, and the terminal lobe of the lower lip the largest. *Calyx* short, five-toothed.

SPECIFIC.—*Inflorescence* a long panicle, the flowers being gathered together in whorls distant from each other. *Stamens* hairy at the base of the filament. *Corolla* violet or sometimes white, having a tuft of white hairs at the base of the inferior lobe of the lower lip. *Leaves* digitate of from five to seven lanceolate lobes, whitish below. *Growth* that of a woody spreading shrub from four to seven feet high. The whole plant has a pungent, aromatic smell, which is especially developed in the fruit.

EXPLANATION OF PLATE XIV.—Plate XIV. represents Vitex Agnus-castus, a plant long considered as an emblem of chastity, from the times even of the Greeks and Romans. Fig. 1, a stamen during the emission of pollen, showing the hairy base of the filament. Fig. 2, a flower. Fig. 3, a cross-section of the fruit, showing the four ovules, one of which is imperfect.

REMARKS.—This natural order is one of the smallest European ones, only including, according to many authors, three or four species from two genera. The best known representative is Verbena officinalis, which, with the plant here figured, was held among the ancients of many parts of Europe as one of the most sacred herbs, and has been offered as a charm for ailments up to a very recent period. None of the plants belonging to this order have a place in British Pharmacopeias. Vitex Agnus-castus grows in two or three other localities along the French shore of the Mediterranean, but becomes more abundant as one moves eastward through Italy, Sicily, and the Levant. It is mentioned as growing in Terai, south of Sikkim, by Dr. Hooker, (Himalaya Journal, vol. i.,

PLATE XIV.

p. 374,) where from the following passage the plant appears to have had odorous white flowers. He says, "In the tropical region the air was scented with the white blossoms of Vitex Agnus-castus, which grew in profusion by the road-side."

The genus Vitex has a wide range, being found in all quarters of the globe ; it is cited from the East Indies, China, Japan, and different parts of India, Ceylon, also from the West Indies, Cayenne, and New Holland. Besides the above stations, three species of this genus have been discovered by Messrs. Speke and Grant in their late African explorations. The Teak tree (Tectona grandis) is a member of this order, and in our gardens the sweet-scented Verbena (Aloysia citriodora) and other representatives are to be recognised.

Vitex Agnus-castus may be found still in blossom in late seasons during the commencement of October. The present specimen was gathered on the shore near Cap Martin, on the 19th of October.

Section of genus Euphorbia, having stipules.

PLATE XV.—*Euphorbia Preslii. Gussone. Woods. Not given by Grenier and Godron.*

GENERIC.—*Styles* 2 to 3. *Capsule* 2 to 3 celled, bursting with elasticity, each cell containing one pendulous ovule. *Male flowers* of one or more stamens. *Perianth* lobed or none. *Flower* usually monœcious.—(Copied nearly from Babington.)

SPECIFIC.—*Capsule* smooth. *Seeds* when ripe somewhat four-sided, brown. *Perianth* furnished with entire milky-white glands, which change as the capsule begins to grow, and soon assume the appearance of the laciniate edges of the perianth itself. *Flower heads* sessile amid bracts, forming terminal clusters. *Stems* smooth, except at the nodes, where they become slightly pubescent, upright, dichotomous. *Leaves* opposite, obliquely-lobed at the base, the deficient side having a margin nearly free from the teeth with which the other edge is surrounded. Both surfaces are more or less hairy, and the short peduncles are pubescent, and expanded below their articulation into minute stipules. Gussone says of the leaves that they are punctate with pellucid spots, "pellucide punctata," but the specimens examined by me bore no trace of this, only showing tiny white patches of white tissue (Cinenchyma) when held up to the light, of which more below. Leaves quickly falling.

EXPLANATION OF PLATE XV.—Plate XV. is of Euphorbia Preslii, a very rare and interesting plant. Fig. 1, a flower-head with the glands present on the perianth. Fig. 2, the same further advanced, when the glands are resolved into the form of the torn edges of the perianth. Fig. 3, the extremity of a flowering branch, showing how the flower heads are sessile amid imbricated bracts. Fig. 4, the completely ripe, four-sided seed. Fig. 5, part of the main stem with one leaf remaining, with its stipule and that of the fallen leaf. It will be noticed that the further half of the leaf is smaller than the other, and is not toothed in its lower part. Fig. 6, the nervature and veining of a leaf, artificially prepared to show it. The intervals between these curious veins are filled with milky tissue, so as to produce, on holding the leaf up to the light, the effect of tiny blotches.

PLATE XV.

REMARKS.—No plant yet found at Mentone can be mistaken for this, though the strange prostrate Euphorbia peplis, which spreads itself over the pebbles just out of the reach of the sea, has the same stipulate leaves, and other curious features in common. Some officinal plants are found in this order, yielding castor oil, which is procured from the seed of Ricinus Palma-christi, a native of India and Africa; Croton oil, from Croton tiglium, a tree growing in India, Ceylon, &c.; and an imitation of the emetic Ipecachuana, which is in reality obtained from a plant of Rubiaceous order, Cephaëlis Ipecachuana. The warmth and equability of the winter climate of Mentone is testified in many ways; the great height (6 or 7 feet) attained by Euphorbia dendroides, and the perennial growth of Ricinus Palma-christi, being worthy to be noticed, as well as the gardens of lemons, and the winter residence of the Rock-martin (Hirundo rupestris).

Besides the drugs mentioned above there are other useful products gained from this curious order of plants, among which may be named tapioca, from Manihot utilissima, and bottle india-rubber, from Siphonia elastica, a native of Brazil.

· The specimen figured was brought from near Ventimiglia, on November 11th. This plant has never yet been found in France, and is not given by Grenier and Godron.

NATURAL ORDER ORCHIDACEÆ.

Tribe—OPHRYDEÆ.

PLATE XVI.—*Serapias cordigera.* *Linn.* *De Candolle.* *Grenier
and Godron.* *Woods.*

GENERIC.—*Column* prolonged into a pointed beak, and bent so as to
be nearly parallel to the lip. *Anther* containing two pollen-masses on
distinct caudicles, which are united in a common gland. *Gland* in a
pouch. *Lip* without a spur, three-lobed, upper lobes bent upwards.
Petals and Sepals united into a helmet.

SPECIFIC.—*Column* terminated by a long beak. *Pollen-masses* dark
green. *Lip* dark brown, having a cordate terminal lobe, densely covered
with curious hairs. *Guiding-plates* (or *Calli*) confluent with the column,
slightly divergent. *Leaves* linear-lanceolate, folded, and spotted like the
lower part of the stem.

EXPLANATION OF PLATE XVI. — Plate XVI. represents Serapias
cordigera. Fig. 1 is of the column, with the guiding-plates conducting
into the Stigmatic-cavity, which last is distinguished by adherent pollen
grains. In this figure the surrounding perianth is cut away, and only
part of the ovary left. Fig. 2, the pollen-masses adhering to part of a
culm of grass, in the position assumed on first withdrawal from
the Anther, bending *away from* the stigmatic-cavity, and separated.
Fig. 3, the same pollen-masses, which have been reversed and drawn
together by some power possessed by the gland. They now bend *towards* the stigmatic-cavity. Fig. 4, Ceratina albilabris, an insect taken
in the interior of a flower of Serapias cordigera, having two pairs of
pollen-masses, from Serapias lingua, fastened on its head. This figure
is of the natural size.

REMARKS.—Mr. Darwin's book on the "Fertilization of Orchids" is
known to many; and not a few of those who have studied it, have,
like myself, found a new and most delightful field for inquiry and
observation. The book shows how in all cases examined by Mr.
Darwin, except that of the self-fertilizing Bee-ophrys, orchidaceous
plants are more or less adapted for insect fertilization, by far the
greater number being totally unable to impregnate their own stigmas
with pollen, but having the pollen-masses fastened to sticky glands,

PLATE XVI.

which glue themselves on to the heads or probosci of different kinds of insects visiting the flowers. We have here a capital example of one of the infinitely varied contrivances by which the pollen-masses, when fastened securely upon some unlucky insect, assume the best direction and position for entering the stigmatic-cavity. By reference to fig. 1, it will be seen how very narrow the stigmatic-cavity is, but that the guiding-plates form a smooth two-walled trough in which the pollen-masses may slide without fail against the stigma. Again, if the pollen-masses were to remain in the position shown at fig. 2, they would both be too wide apart to enter the stigmatic-cavity, and be turned away from it. By a wonderful power connected with the contraction of the gland, these pollen-masses are drawn together as they descend, always moving in the right direction, and not only overcoming their natural weight, but also traversing two planes. The little insect drawn at fig. 4 was taken by me within the helmet of Serapias cordigera, having left grains of yellow pollen from the masses on its head on the stigma of that flower; the strange pollen being easily distinguished as belonging to Serapias lingua by its yellow colour. I do not feel sure of the complete distinctness of S. cordigera as a species; at any rate, there are many links formed by hybridisation which may lead one to doubt what characters are most peculiar to it. Parlatore makes a species of S. neglecta, while there are some who would unite S. cordigera and S. longipetala. I hope to gather more facts on this most interesting genus, which I may relate in Part II. of the present volume, when I figure two other species of Serapias. These plants abound at Mentone, coming into flower about the end of April. The specimen figured was gathered on Montegrosso, May 11th.

NATURAL ORDER ORCHIDACEÆ.

Tribe—OPHRYDEÆ. Section of genus Orchis, having united glands.

PLATE XVII.—*Orchis longibracteata. Bivona. De Candolle. Woods. Aceras longibracteata. Grenier & Godron. Barlia longibracteata. Parlatore. Loroglossum longibracteatum. Ardoino.*

GENERIC.—*Perianth* ringent, hooded. *Lip* 3-lobed, spurred. *Glands* of the stalks of the Pollen-masses in a common pouch.—Babington's Manual.

SPECIFIC.—*Stigmatic chamber* vertical, pear-shaped, unusually large and distinct. *Pollen-masses* having their stalks united in a common gland. *Staminodia* distinct. *Lip* divided into three lobes, the terminal lobe being notched, and the lateral ones wavy and of variable length ; colour very inconstant, but frequently purplish with dusky edges. *Spur* very short. *Petals* lanceolate, enclosed by the hood-shaped upper sepal. *Sepals* veined, the two lateral ones spotted with pinkinside. *Bracts* longer than the flowers. *Leaves* very large, frequently ten and sometimes thirteen inches long, by from three to five inches broad. *Tubers* large undivided.

EXPLANATION OF PLATE XVII.—In this plate is drawn Orchis longibracteata, with the large insect (Xylocopa violacea) which sometimes visits these flowers, and bears away their pollen-masses. Fig. 1, Pollen-masses adhering by their flat, viscid gland to a bit of culm of grass, and widely separated as they appear on first removal. Fig. 2, the same, but having been exposed about 1¼ minute to the air, the Pollen-masses approach one another, still remaining nearly upright. Fig. 3, the same in the third position, bringing the stalks nearly parallel to the surface of the culm of grass. Fig. 4, a flower with 2 sepals, one petal and three quarters of the lip taken away ; the long bract is seen clinging round the base of the ovary, and the pear-shaped stigmatic chamber is represented as having been touched with pollen, grains of which are seen adhering on its surface ; above this chamber the pouch projects, standing high above the surface of the lip, and on either side of the Anther, glands or staminodia are found. These glands (staminodia) indicate the position of the Anthers in Cypripedium. Fig. 5, a section of the spur cut parallel to the column, showing a curious fold in the interior which is

PLATE XVII.

covered by hair-like processes. Fig. 6, Xylocopa violacea with the pollen-masses of Orchis longibracteata attached between its eyes. This figure is of the natural size.

REMARKS.—This fine plant stands quite at the head of European orchids in respect of size and time of flowering, and has no very immediate aspirant for supremacy ; the nearest species being Orchis hircina, which, though most interesting and remarkable, can in no way claim either stateliness or beauty. The movements of the pollen-masses are also very striking, as the traversing of either plane (*i.e.*, the plane of contraction and the plane of depression) is distinctly performed, the depression not setting in till after the masses are drawn close together. By way of comparison it is interesting to examine Orchis pyramidalis, which has the greatest similarity in action of the pollen-masses of any mentioned by Mr. Darwin. In this case the pollen-masses are fastened to a gland, which, when taking hold of the object that removes it, curls up, and separates the masses, thus enabling them to strike two widely-divided stigmas. In the case of Serapias, Orchis hircina and Orchis longibracteata, the reverse action takes place, without any resemblance in the gland, which remains flat and apparently unchanged. The height at which the pouch stands seems to me a point always worthy of notice, as the position is, I believe, relative to the size and make of the insect best qualified to remove the pollen-masses. In this very case if the pouch, to depress which a hard push is necessary, had been lower down, the great Bee (Xylocopa violacea) could not have struck against it with his head, and a smaller insect, or the proboscis of Xylocopa, would probably not have given a sufficient blow. I was favoured with a foreign specimen of Orchis hircina by the kindness of friends, after my return to England ; and I found that that most extraordinary plant has much smaller pollen-masses, and a pouch placed quite low on the column, so that the gland could not possibly be removed in the same way as that of Orchis longibracteata. The geographical range, given by Parlatore in his Flora Italiana, is especially traceable round the upper curve of the Mediterranean shores, extending as far as Spain to the west, and Crete and Scio to the east. It grows northward as far as Arles, and southward to Algeria and the Canary Islands. I find it also cited as growing on mount Hymettus, and the hills round Athens. Orchis longibracteata comes into flower about Christmas, and continues in blow until the end of March. The specimen drawn was brought from the Mentone Valley, where this plant abounds on rough banks under pine-trees.

Tribe—OPHRYDEÆ. Section of genus Orchis, having two separate glands, and the lateral sepals spreading.

PLATE XVIII.—*Orchis Olbiensis ?* This name was given me for the present plant by M. Ardoino, but he was unable to cite the work in which he believed that it was described. It is not mentioned by any author whose works I have.

GENERIC.—See description of preceding plate.

SPECIFIC.—*Stigmatic chamber* much below the level of the lip. *Pollen-masses* dark green. *Staminodia* distinct. *Lip* deeply cut into three acuminate lobes ; each lateral lobe has generally acuminate teeth on the margin, and the terminal lobe a central tooth and two pairs of lateral teeth. The colour of the lip, and of the whole flower indeed, is pink, though on its surface some tiny hairlike processes give the appearance of brown spots. *Spur* as long or longer than the ovary, generally horizontal. *Petals* ovate, short, forming a hood over the column. *Sepals* lanceolate-acuminate, the central one not bent over the column. *Bract* acuminate, coloured, about the length of the ovary. *Leaves* broadly linear-lanceolate, obtuse, mucronate, erect, not spreading on the ground.

EXPLANATION OF PLATE XVIII.—Orchis olbiensis, if that be indeed its name, here figured, is one of the many examples of European plants about which very little is known, or at any rate reported. Figs. 1 and 2 are of the pollen-masses performing the necessary movements after removal. Fig. 2 shows one of the pollen-masses immediately after removal by the culm of grass to which it is attached, when it is perfectly upright. Fig. 2 represents it when prostrated and in the proper position for entering the stigmatic-cavity. Fig. 3 is of the column and part of the spur, placed in such a position that the stigmatic-cavity is seen underneath the anther with the grains of pollen adherent in it. Fig. 4, an entire flower explaining the relative parts.

REMARKS.—Orchis olbiensis is said to be the name given to a plant similar to this which was discovered at Hyères (once called Olbia) and M. Ardoino considered that this plant, on its discovery in 1863 on the Aiguille Mountain, at Mentone, was probably identical with it. This Orchis varies much in size, and specimens may be found growing to

PLATE XVIII.

twice the height of those here drawn. I believe that the colour, the finely-pointed ends of the flowers, and the upright leaves are tolerably constant and effective distinctions. The mode of fertilization is simple, so that when the pollen-masses are once removed by insects, each mass has only to be bent forwards without any lateral sweep. In some Orchids, similar to this in many respects, but which have two separated stigmas, one on each side of the pouch, the pollen-masses swerve to the right and to the left, each aiming at the stigma nearest to its own side of the anther. This is seen in Orchis ustulata and others. Orchis olbiensis has only yet been found at Mentoue on the higher mountains, to the summits of which it reaches. The Aiguille and Mulaciers mountains are its chief habitats, and the present known range is from about 2,500 to about 4,000 feet. Perhaps this plant may be placed between Orchis provincialis and O. laxiflora, having some likeness to O. acuminata and O. Haurii of Jordan, though not coinciding with either. Orchis olbiensis flowers from about the end of April to the middle of May. The specimens drawn in the plate were brought me from the Mulaciers mountain on May 4th.

Tribe—OPHRYDEÆ. Section of genus Ophrys, having the lip
three-lobed, and the posterior lobe largest.

PLATE XIX.—*Ophrys scolopax. Cavanilles. Grenier and Godron.
Not given by Woods.*

GENERIC.—*Pollen-masses* having two separate glands, enclosed in two
pouches which appear distinct. *Ovary* not twisted. *Lip* thick, rather
fleshy, without any spur.

SPECIFIC.—*Lip* deeply three-lobed, the terminal lobe having at its
extremity a three-pointed, upturned process. The margins of the lip
are frequently much reflexed.

EXPLANATION OF PLATE XIX.—Ophrys scolopax, the plant drawn in
this plate deserves most particular attention, as the facts given below
attest. Both of the spikes represented in this plate are entirely unable to
fertilize themselves. Fig. 1, a flower taken from a spike gathered at
Cannes, and in the act of self-fertilization, the pollen-masses bending
into the stigmatic-cavity while the glands remain in the pouch.
Fig. 2, part of a flower taken from a spike gathered at Mentone, with
the pollen-masses still in the anther-cells, and the stigma fertilized.

REMARKS.—I am able to show in the present plate that an Ophrys
exists which presents two forms, growing in two widely-separated places.
The facts stand thus : all the specimens of Ophrys Scolopax which I
was able to procure at Mentone were entirely unable to fertilize themselves, being, like all the other representatives of the genus Ophrys
known to me there, except the rare Bee Ophrys (O. apifera.), quite without the power of releasing the pollen-masses from the Anther-cells, and
bending them into the stigmatic-chamber. While at Mentone, therefore, I set down O. scolopax as always indebted to insects for its
fertilization. When passing through Cannes, however, some kind friends
sent me a bundle of the Orchids of that place, among which I found a
bunch of Ophrys scolopax. How great was my surprise to find the
flowers all fertilizing themselves, the pollen-masses being bent down into
the stigmatic-cavity as in the drawing at fig 1. It is curious, as a
coincidence if in no other way, that at Cannes the Bee Ophrys is as
abundant as it is scarce at Mentone ; so that self-fertilization is the rule

PLATE XIX.

at the former, and the exception at the latter place. I have neither facts nor experience enough as yet to justify my attempting any induction; it is possible, however, that if I am able to complete the second part of this volume after another year's attentive observation, I may append some further notice of this plant when describing the beautiful yellow Ophrys (O. lutea) and some other handsome representatives of the genus found at Mentone. I venture to suggest the following queries as subjects for investigation especially bearing on the present remarks to those who are interested in the " Fertilization of Orchids." 1st, Does Ophrys scolopax exist at Mentone in a condition of self-fertilization? 2nd, Is Ophrys apifera ever indebted to insects for its fertilization? 3rd, What are the habits of Ophrys arachnites? The specimens figured were brought me on April 22nd, when this plant is in full flower. The habitats chosen by Ophrys scolopax are chiefly on rough banks such as are found in both the Eastern and Western bays. I so much doubt whether this plant can be really called a species, that I am unwilling to commit myself by saying that it can always be distinguished from any other Ophrys growing at Mentone.

Pl.16

NATURAL ORDER IRIDACEÆ.

PLATE XX.—*Crocus medius. Balbis. Parlatore. Woods. Not given by Grenier and Godron.*

GENERIC.—*Style* very long, filiform, having a broad three-lobed stigma, the margin of which is generally cut more or less. *Perianth* regular, campanulate, with a very long tube.

SPECIFIC.—*Stigmas* cut into a spreading head of capillary segments, standing higher than the anthers. *Anthers* much longer than the filaments. *Flower* solitary, purple, autumnal, appearing without any leaves, having the divisions of the perianth smooth at the throat, and much shorter than the tube. *Leaves* broadly linear, erect, the slight channelling forming a white line on the upper surface. *Bulb* subglobose, covered with a reticulate tunic.

EXPLANATION OF PLATE XX.—Plate XX. represents Crocus medius, the flowers of which were gathered in November, and the leaves and capsule in April. Fig. 1, the pistil, bearing a multifid stigma, and having the lower extremity clasped by the tube, which is shown divided in half. Fig. 2, the netted (reticulate) covering of the bulb.

REMARKS.—According to Parlatore, this beautiful autumnal Crocus is peculiar to the western part of the Italian peninsula; and the following statement, made in the " Botanical Register" (Bot. Reg., 1845, pl. 37, fig. 5), where a drawing is given, gives additional confirmation :—" Crocus medius grows in the mountain meadows near Varese in Liguria, and in some parts of the Riviera of Genoa. It was named by Balbis as intermediate between Crocus sativus and C. pyrenaicus ; but it is, in fact, a link between C. byzantinus and C. pyrenaicus." Crocus medius is very rare at Mentone, and the English visitors are scarcely arrived before the flowering is nearly at an end. The specimens figured were gathered in the Western Bay on November 6th, when the flowers were very scarce. This plant is one of the very few species of Crocus which have the same habit as Colchicum, and produce their blossoms so late in the autumn that the capsules are obliged to lie dormant throughout the winter waiting for the influence of the spring sun, when they appear with the leaves. No doubt there is some very good reason why the leaves desert the flower and reserve their energies in order to assist at the ripening of

PLATE XX.

the fruit. There is no other species of Crocus yet discovered near Mentone with which the present one can be confused ; indeed, the only other representative of the genus found in the immediate neighbourhood is the beautiful C. versicolor, which flowers with its leaves in the spring. Crocus medius flowers in October, and specimens may still be procured in the early part of November. The capsule and leaves were gathered for me in April.

NATURAL ORDER AMARYLLIDACEÆ.

PLATE XXI.—*Leucoium hiemale. Woods. De Candolle, variety a. Ruminia hiemalis. Parlatore.*

GENERIC.—*Stigma* simple, stamens 6, having short filaments inserted on an epigynous disc. *Anthers* bilocular, either cell opening from above by a longitudinal slit. *Perianth* campanulate, not narrowed into a long tube.

SPECIFIC.—*Stigma* papillose, obtuse. *Style* erect, filiform. *Capsule* pear-shaped. *Seeds* shiny-black, having a white caruncle extending along from the hilum to the opposite extremity. *Stamens* equal, connivent, having their filaments bent angularly and inserted on the 6-lobed epigynous disc. *Anthers* oblong, introrse, opening at the top, adhering to one another by their bases. *Divisions of perianth* white, the inner segments rather shorter and obtuse, the outer pointed with thickened tips. *Spathe* of two valves longer than the pedicels. *Peduncle* strong, generally but one from each bulb, sometimes two. *Leaves* broadly linear, sometimes appearing long before the flowers.

EXPLANATION OF PLATE XXI.—Plate XXI. is, I believe, the first published drawing of Leucoium hiemale. Fig. 1 shows the filiform style. Fig. 2, a seed with its caruncle. Fig. 3, a stamen, showing the angle in the filament and turned outwards so as to bring the dehiscence into view. Fig. 4, a flower with the three nearest divisions of the perianth taken away, displaying the connivent anthers with their filaments inserted on the curiously-lobed disc.

REMARKS.—Among the mysteries which surround the origin of all vegetable life, that of the limitation of species to certain districts seems not the least incomprehensible. Here, for instance, is a species of Leucoium which is believed to have but one habitat on the face of the earth, claiming only a small strip of rocky shore reaching from Nice to about two miles east of Mentone. Leucoium hiemale grows in a stony soil, and out of the cracks of the hardest limestone rocks at Pont St. Louis, Capo Veglio on the way to Monaco, and at some height on the Aggel Mountain, besides other less abundant localities. Neither Galan-

PLATE XXI.

thus nor Leucoium vernum have been found nearer than the Sospello valley, and there is no plant growing at Mentone which at all resembles this. From its name one might expect that Leucoium hiemale flowered in the winter, though April is the month when it is most fully in blow, and I never have been able to find a bud even before March. The specimens figured were gathered at Pont St. Louis.

NATURAL ORDER AMARYLLIDACEÆ.

Tribe—NARCISSEÆ.   Section of genus Narcissus, having yellow flowers
and flattened leaves.

PLATE XXII.—*Narcissus aureus.   Loiseleur.   Grenier and Godron.
Parlatore.   Narcissus chrysanthos Woods, part.*

GENERIC.—*Stamens* inserted either in the crown or base of the tube
of the corolla.   *Perianth* of six spreading divisions, within which is a
bell-shaped crown.

SPECIFIC.—*Divisions of perianth* very broad in proportion to their
length ; the three outer pointed with a woolly tip.   These divisions are
bright yellow and often about thrice as long as the orange-coloured
crown.   *Leaves* broadly linear, slightly channeled.   *Stem* sub-cylindrical,
of variable height, from one foot six inches to two feet from the ground.

DESCRIPTION OF PLATE XXII.—PLATE XXII. represents Narcissus
aureus, which is apparently identical with the soleil d'or of our gardens.

REMARKS.—Whether this beautiful Narcissus may claim to be a
native of Italy, or even of Europe, I do not know ; I am only able to
state that it now grows in some places where there is no likelihood of its
having been planted.   There is no Mentonese plant known to me with
which the present can be confused if colour be allowed to be a specific
character ; there may be, however, one or two of those pale yellow
Narcissi which some consider as distinct species.   The only habitat
where this plant was known as growing in France by Grenier and
Godron was Grasse, and it is always considered a rare plant by foreign
botanists.   The specimen from which I have drawn was gathered in
the Eastern Bay on January 26th.   The flowering commences in
January and ends in February, varying according to the season.

NATURAL ORDER AMARYLLIDACEÆ.

Tribe.—NARCISSEÆ.   Section of genus Narcissus, having yellow
flowers and flattened leaves.

PLATE XXIII.—*Narcissus tazetta.   Linn.   De Candolle.
Grenier and Godron.   Woods.*

GENERIC.—*See preceding description.*

SPECIFIC.—*Flowers* about eight in number, always having pale, yellow-
tinted divisions and a bright yellow crown.   The shape and relative
length of the parts of each flower vary much on the same scape.   *Scape*
much shorter than in other Mentonese Narcissi, being generally less
than one foot in height.   *Leaves* of very variable width, but nearly of
the same height as the scape.

EXPLANATION OF PLATE XXIII.—Plate XXIII. represents Narcissus
tazetta.

REMARKS.—This is the commonest of all the representatives of the
genus Narcissus about Mentone, and is one of the very few species
about which no doubt exists as to its title to be called a native.   The
profusion in which this lovely plant grows in the olive and vineyards
in the neighbourhood can only be compared with the abundance
of our own daffodils, growing in the well-remembered English nooks.
The daffodil is not found at Mentone, though we possess Narcissus
incomparabilis, which must be considered as closely connected with it.
Narcissus tazetta varies much in the shape of the flowers and leaves,
and these features are considered by some as worthy to rank as specific
characters, otherwise no plant known to me at Mentone can be mistaken for this.   The specimen drawn was gathered at the Palazzo
Orenga, in a great open flat partly under cultivation, on February 29th.
The flowering begins in February and ends in March, varying according
to the season.

Tribe—TULIPEÆ. Section of genus Tulipa, having the filaments of the stamens smooth.

PLATE XXIV.—*Tulipa clusiana. De Candolle. Woods. Grenier and Godron.*

GENERIC.—*Style* wanting. *Stigma* 3-lobed. *Stamens* erect. *Divisions of Perianth* free, without nectaries. *Seeds* flat.

SPECIFIC.—*Stamens* longer than ovary. *Divisions of Perianth*, the outer lanceolate, the inner elliptic obtuse. The colour as in the plate. *Leaves* glaucous, linear or linear-lanceolate. *Bulb* small, not woolly.

EXPLANATION OF PLATE XXIV.—Plate XXIV. represents Tulipa clusiana.

REMARKS.—It was not until I had seen flowers fully expanded in their wild state, that I ventured to draw the original sketch for the present plate, as I am perfectly aware that it is the exception to find them advanced beyond the apparent condition of buds. I found that the blossoms of Tulipa clusiana were quickly affected by the increase or diminution of direct sun-light, and I was only able to make the present drawing by sitting in the full blaze of a bright sun, under the influence of which the flowers rapidly expanded, assuming on withdrawal the more familiar appearance represented in the upper part of the plate. It is possible to make this Tulip open and shut its blossoms many times by subjecting it to the above conditions, and I am quite at a loss to conjecture what advantage is gained by the plant from this most curious sensibility.

Tulipa clusiana is found in the Olive and Lemon terraces near Mentone, both in eastern and western bays, but it is more especially abundant in the Latte valley between Mentone and Ventimiglia, where the fine specimens figured in this plate were gathered. The one other Tulip (T. præcox) known as growing at Mentone is scarlet and black, besides being otherwise amply distinguished from the present plant. The time of flowering is from the end of March to the middle of April.

NATURAL ORDER LILACEÆ.

PLATE XXV.—*Frittilaria delphinensis. Grenier ?* Grenier
and Godron. *Not given by Woods.*

GENERIC.—*Style* trifid. *Seeds* flat. *Perianth* of six deciduous divisions, having nectariferous depressions.

SPECIFIC.—*Style* becoming gradually wider from the base up, generally deeply trifid. In the description of Frittilaria delphinensis given by Grenier and Godron, the divisions of the style are said to be very short. *Stigmas* papillose, channeled. *Ovary* longer than broad. *Divisions of Perianth* yellow, sparingly chequered with reddish-brown ; the exterior oblong, small, the interior broadly oboval, much larger. Grenier and Godron give "purple-brown, rarely yellow," as the colour of the divisions. *Flower* nearly as broad as long. *Leaves* generally four or five, upright, lanceolate, clustered together near the flower.

EXPLANATION OF PLATE XXV.—I believe this to be the first coloured plate of Frittilaria delphinensis. Fig. 1 is of a very deeply cleft style, drawn from a specimen in which this feature was especially marked. Fig. 2, three of the stamens which were not removed when the divisions of the perianth were taken away, with the ovary and style, the summit of which is less divided. Fig. 3, an outer and inner division of the perianth of the natural size.

REMARKS.—I have little doubt that the plant here figured is indeed Frittilaria delphinensis of Grenier, though in some respects my specimens did not answer to his description. In all the flowers which I have examined the yellow colour predominates, and the style is deeply divided, whereas Grenier found them generally of a purple-brown colour, with styles but slightly divided. It must be remembered that he describes plants from very different and distant habitats, as may be seen by the following list of stations :—"Hautes alpes du Dauphiné, environs de Gap, Glaise, Séuse, etc. ; l'Arche, Monte-Viso, Lautaret, Lusette en Luz, dans la Drôme." It is also to be noticed that he gives August as its time of flowering. It is probable, therefore, that the different climate and season under which my specimens were found growing is sufficient to account for some variation. Frittilaria

PLATE XXV.

delphinensis has not yet been discovered on the mountains near Mentone, and comes from a mountain not far from Giandola in the Sospello valley, at about four thousand feet elevation. The only Frittilary known as yet at Mentone is F. involucrata, which, besides having a whorl of three leaves above, and opposite leaves below, is distinguished at a glance by its very graceful pliant stem and small dark flowers. Frittilaria delphinensis was in full blow on April 23rd, when I received the specimens figured, from the above-mentioned locality.

MARY S. RICKERBY, Printer, Bell Court, 4A, Walbrook, London. E. C.

PLATE XXVI.

HYPECOUM PROCUMBENS, Linn.

Natural Order PAPAVERACEÆ.

GEN. CHAR.—*Sepals* 2. *Petals* 4. *Stamens* 4, opposite the petals *Style* bifid, *lobes* subulate. *Capsule* linear, divided by septa, either lomentaceous or dehiscing by two valves, bearing the placentas on the margin (Hook. et Benth. Gen. Plant. i. 54). *Plants* with watery juice.

SPEC. CHAR.—*Sepals* 2, oval, equalling about a third of the length of the corolla. *Petals* irregular, the two exterior large and three-lobed above, the two interior smaller, trifid ("rarely entire or bifid," Gren. et Godr.), the central lobe fringed. *Capsule* lomentaceous, bent.

Hypecoum procumbens, Linn. Sp. Plant. p. 181; Gren. et Godr. Fl. de Fr. i. 62; Woods, Tour. Fl. p. 12.

HABITAT.—Cultivated ground on sandstone formation. February, March.

REMARKS.—In the 'Genera Plantarum' of Messrs. Bentham and Hooker (London, 1862), Hypecoum is placed at the commencement of the suborder Fumarieæ; the Fumitories being considered by them as belonging to the Natural Order Papaveraceæ. Thus, Hypecoum stands with Corydalis on the one hand, and Eschscholtzia on the other. The geographical range of the genus is thus described (p. 54):—"There are 4 (or 5?) species growing in Southern Europe, Northern Africa, and the temperate regions of Asia." The structure of the two inner petals is very curious, and probably has reference to the fertilization of the plant. Before the opening of the flower, the fringed middle lobe of each of these petals is closely wrapped round the stamens, so that on the expansion of the flower the pollen is wiped out of the cells by the motion of withdrawal. As soon as the petals are spread, the fringed margins of this middle lobe become recurved, and expose the pollen so that the bodies of insects visiting the flowers may readily come in contact with it. The plants figured were gathered on terraces near the cemetery in March, and are probably intermediate between Hypecoum procumbens, Linn., and H. grandiflorum, Benth. (Cat. Pl. Pyr. 91). This latter species is now abandoned by its author.

EXPLANATION OF PLATE XXVI.—Fig. 1, an entire flower. Fig. 2, an exterior petal. Fig. 3, an interior petal from a bud. Fig. 4, the same, with the middle lobe recurved. Fig. 5, stamens and pistil. All the figures are magnified.

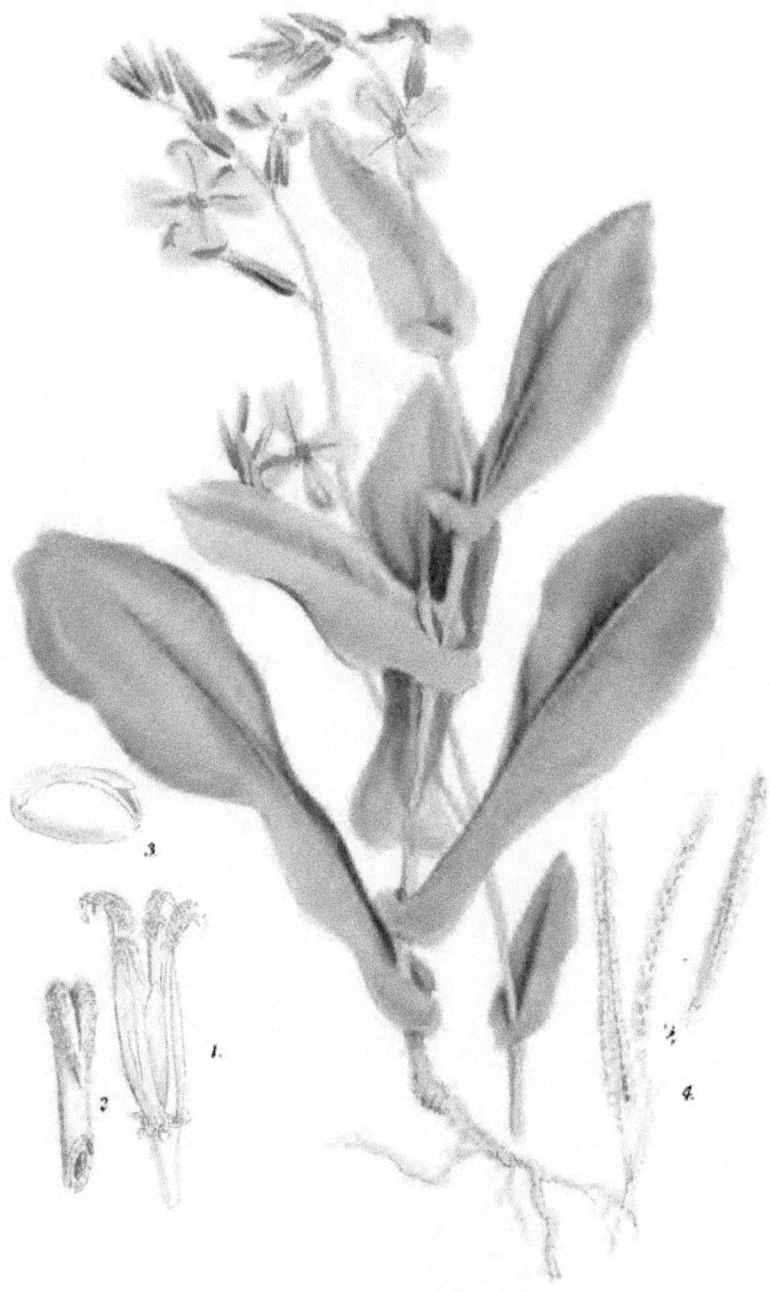

PLATE XXVII.

MORICANDIA arvensis, DC.

Natural Order CRUCIFERÆ.

GEN. CHAR.—*Calyx* having two of the sepals saccate. *Petals* purple or pink. *Stigmas* confluent. *Siliqua* one-nerved. *Seeds* in one or two rows ; *cotyledons* conduplicate.

SPEC. CHAR.—*Sepals* longer than the pedicel. *Stigmas* bilobed. *Siliqua* keeled. *Leaves* glaucous, amplexicaul. *Stem* often woody. *Growth* biennial.

Moricandia arvensis, DC. Syst. ii. 626 ; Gren. et Godr. Fl. de Fr. i. 82 ; Woods, Tour. Fl. p. 25.

HABITAT.—Clay of conglomerate near Ventimiglia, nearly from shore-level to 1000 feet elevation. February to April.

REMARKS.—The genus Moricandia contains five species, which grow in the south of Europe, Northern Africa, and Western Asia (Hook. and Benth. Gen. Pl. i. 85). It is curious that the present species is so very rare a European plant, considering the profuse fertility of its pods and the readiness with which it comes from seed. The plants figured were gathered at Ventimiglia in February, where flowers may be found throughout the winter. Moricandia arvensis grows at Marseilles and in Sicily, but is rare at the former place.

EXPLANATION OF PLATE XXVII.—Fig. 1, a flower deprived of its petals and sepals. Fig. 2, the bilobed stigma. Fig. 3, the conduplicate cotyledons. Fig. 4, the ripe siliquas. All the figures, excepting the last, are magnified.

1.

2.

PLATE XXVIII.

CYTISUS HIRSUTUS, Linn.

Natural Order LEGUMINOSÆ.

Section of genus Cytisus having long tubular calyces.

GEN. CHAR.—*Calyx* persistent, 2-lipped; lips short, oval, divergent; the upper lip truncate or bifid; the lower three-toothed. *Stigma* oblique. *Pod* linear-oblong, compressed, greatly exceeding the calyx; *cotyledons* leafy in germination.—Trees or shrubs without thorns; leaves trifoliate, rarely unifoliate. Gren. et Godr. Fl. de Fr. i. 358.

SPEC. CHAR.—*Flowers* springing from the nodes; pedicels about one-third of calyx. *Calyx* covered with spreading hairs; lobes of upper lip obtuse. *Standard* obtuse, nearly orbicular. *Pod* hairy, nearly flat. *Leaves* trifoliate or variously reduced by abortion, hairy. *Stems* hairy, erect, or ascending, woody.

Cytisus hirsutus, Linn. Sp. Plant. 1042; Gren. et Godr. Fl. de Fr. i. 361; Woods, Tour. Fl. p. 79.

HABITAT.—Turin Valley, on sandstone. April.

REMARKS.—Cytisus hirsutus, Linn., is perhaps the rarest Leguminous plant to be found at Mentone, as it appears that only one French habitat was known to MM. Grenier and Godron when they published their 'Flore de France' in 1848. This station was among the Alps of Dauphiné.*

EXPLANATION OF PLATE XXVIII.—Fig. 1, the calyx and the reproductive organs, magnified. Fig. 2, a pod, of the natural size.

---

* It is also found near Genoa and Lugano, and is said by De Candolle (Prodromus) to range from the former place to Hungary.

Vincen

PLATE XXIX.

CAMPANULA MACRORHIZA.

Natural Order CAMPANULACEÆ.

Section of the genus Campanula having the capsules erect.

GEN. CHAR.—*Calyx* 5-parted. *Corolla* bell-shaped or rotate, 5-lobed. *Stamens* 5, free. *Style* cleft into 3–5 stigmatic lobes. *Capsule* turbinate, 3–5-celled, cells opening by pores.

SPEC. CHAR.—*Calyx-segments* reflexed or sometimes spreading, acuminate. *Capsule* erect. *Root* thick, woody.

Campanula macrorhiza, Gay in DC. Prod. vii. 475 ; Woods, Tour. Fl. p. 239.

HABITAT.—Clefts of Jurassic limestone, from shore to 4000 feet elevation. January to April.

REMARKS.—Campanula macrorhiza, Gay, takes the place of Campanula rotundifolia, L., in this department, and I have never yet heard of these plants being found in the same locality. The distinctions given above as separating this plant as a species are but slight, and it is quite possible that intermediates may be found in which the pendent capsule of C. rotundifolia, Linn., occurs on plants having the reflexed calyx-segments of C. macrorhiza, Gay. I have, however, been unable to find any such intermediates, either growing or in herbaria. The specimens figured were gathered at Pont St. Louis in February, 1864.

EXPLANATION OF PLATE XXIX.—Fig. 1, mature capsules of the natural size.

PLATE XXX.

CONVOLVULUS ALTHÆOIDES, Linn.

Natural Order CONVOLVULACEÆ.

Section of genus Convolvulus having a twining stem.

GEN. CHAR.—*Corolla* campanulate, having 5 angles and 5 folds. *Style* filiform ; *stigmas* 2. *Capsule* 2–4-celled, each cell containing 1–2 seeds.

SPEC. CHAR.—*Flowers* solitary or in pairs, on long axillary peduncles ; *bracts* linear, distant from the calyx. " *Capsule* ovoid, glabrous " (Gren. et Grodr.). *Leaves* more or less cordate, entire or variously lobed. *Stem* generally hairy, as is the whole plant ; sometimes scarcely twining.

Convolvulus althæoides, Linn. Sp. Plant, p. 222 ; Gren. et Godr. ii. 501 ; Woods, Tour. Fl. p. 250.

HABITAT.—Dry banks among the olive terraces. April, May.

REMARKS.—As far as my observations go, this plant is pretty constant at Mentone to the form under which I have represented it ; but at Cannes it sometimes appears with silky leaves and much smaller flowers. This latter form is treated by MM. Grenier and Godron as a variety ; but as a species by De Candolle (Fl. Fr. Suppl. p. 423), who calls it C. argyræus. Convolvulus althæoides, Linn., has a wide range, and is mentioned by De Candolle (Geog. Bot. i. 409) as reaching from Madeira, the Canaries, and Mogador, round the Mediterranean shores as far as Egypt and Anatolia, but not growing near the Black Sea or Caucasian range. The silky variety mentioned above is cited by Koch (Syn. Fl. Germ. et Helv. ii. 429) as growing in Istria, and is called C. tenuissimus, Sibth. and Smith (Fl. Græc. Prod. i. 134), a name adopted by Woods (Tour. Fl. p. 250), who joins the above-named authors in considering it as a species.

PLATE XXXI.

ECHIUM CALYCINUM, Viv.

Natural Order BORAGINEÆ.

Section of genus Echium in which the lower flowers do not spring from the axils of the bracts.

GEN. CHAR.—*Corolla* irregular; *throat* naked; *limb* oblique, of 5 unequal lobes. *Stamens* unequal. *Achenes* 4.

SPEC. CHAR.—*Bracts* oblong or lanceolate, equal at the base. *Calyx* developing at the ripening of the fruit, so as to become leavy. *Stamens* included. *Leaves* hispid, tubercular, oblong or obovate, the lower ones narrowed into a petiole, the upper sessile. *Stems* prostrate or ascending, hispid and tubercular. Gren. et Godr. Fl. de Fr. ii. 525.

Echium calycinum, Viv. Ann. Bot. i. pars. 2, p. 164, and Fl. It. Fragm. i. p. 2, tab. 4; Gren. et Godr. Fl. de Fr. ii. 525; Woods, Tour. Fl. p. 256.

HABITAT.—Stony ground on Jurassic limestone, near the shore. March.

REMARKS.—This plant is rather local at Mentone, but may be found in abundance above Pont St. Louis, on the western side of the gorge.* The specimens figured were gathered there in the end of March, 1865.

EXPLANATION OF PLATE XXXI.—Fig. 1, an entire flower. Fig. 2, the stamens and style, with part of the corolla and calyx. Both figures magnified.

---

* Echium calycinum, Viv., grows near the sea, at intervals, between Marseilles and Naples; is found at La Ciotat, on the Ile Sainte-Marguerite opposite Cannes, and in Sicily.

PLATE XXXII.

MICROMERIA PIPERELLA, Benth.

Natural Order LABIATÆ.

GEN. CHAR.—" *Calyx* cylindrical, 13- rarely about 15-ribbed ; *teeth* nearly equal, straight or scarcely forming two lips; *throat* generally hairy. *Corolla* straight ; *limb* bilabiate ; *upper lip* erect, nearly flat, entire or emarginate. *Stamens* 4, ascending, approximate above or more rarely somewhat divergent; *anthers* free, 2-celled ; *cells* distinct. *Achenes* dry, smooth." Benth. Lab. et Scroph. p. 369.

SPEC. CHAR.—*Fasciculi* few-flowered, the common *peduncle* nearly as long as the floral leaves. *Calyx* tubular, pubescent, 13-ribbed ; *teeth* subulate; *throat* hairy within. *Corolla* pubescent, twice the calyx in length. *Leaves* ovate, sometimes cordate at the base.

Micromeria piperella, Benth. in DC. Prodr. xii. 221 ; Woods, Tour. Fl. p. 288.

HABITAT.—Mountains near Mentone, at not less than 3000 feet elevation. September, October.

REMARKS.—The plant here figured is peculiar to the maritime Alps, where it is tolerably abundant at high elevations. Those who are at Mentone as early as the commencement of October may find M. piperella, Benth., in its greatest beauty ; but to the greater number of visitors the flowers are but little known, on account of the lateness of their arrival. The most distinctive feature of this genus is the calyx, which separates it from Thymus by its nearly equal teeth, which are not disposed in two lips; from Satureia by its thirteen instead of ten nerves; and from Calamintha by its greater regularity (Benth. in DC. Prodr. xii. 212). The division of Labiatæ into tribes, founded upon the position of the stamens, seems to me a little difficult of apprehension, as it sometimes happens that in the same species two forms exist, in one of which the character of a tribe not its own is found. This seems to be the case here, as the anthers in M. piperella, instead of being convergent under the upper lip of the corolla, are widely separated, as in the tribe Thymeæ.

EXPLANATION OF PLATE XXXII.—Fig. 1, a flower and bracts. Fig. 2, the corolla. Figs. 3 and 4, the front and back of a stamen. Fig. 5, stigma and part of style. Fig. 6, the calyx (drawn as seen against the light), showing the thirteen nerves, two of which branch.

PLATE XXXIII.

STATICE PUBESCENS, DC.

Natural Order PLUMBAGINEÆ.

Section of genus Statice having the middle and lower branches sterile.

GEN. CHAR.—*Flowers* scattered. *Calyx* scarious, with a coloured nerve. *Petals* united at their base or sometimes free. *Stamens* inserted at the base of the petals. *Styles* glabrous, free or united only at the base.

SPEC. CHAR.—*Flowers* on the uppermost branches; *spikelets* 1–2-flowered, straight; *bracts* pubescent, the outer about a quarter of the inner, scarious and orbicular, the inner scarious only at the edge, *Calyx* straight, pubescent. *Leaves* coriaceous, of uniform texture, densely pubescent, spathulate, sometimes emarginate; *margins* revolute. *Scapes* straight, slender, pubescent.

Statice pubescens, DC. Fl. Fr. v. 380; Gren. et Godr. Fl. de Fr. ii. 748; Woods, Tour. Fl. p. 306.

HABITAT.—Rocks close to the sea, Cap Martin, etc. October.

REMARKS.—In the immediate neighbourhood of Mentone we have but three representatives of the Natural Order Plumbagineæ, namely, Armeria plantaginea, Willd. (Enum. Hort. Berol. i. 334), Plumbago europæa, Linn. (Sp. Plant. p. 215), and Statice pubescens, DC. The first of these grows at an elevation of not less than 3200 feet (Ardoino, Cat. des Pl. Vasc. de Menton, etc.), and the second on warm rocky and stony ground at a hundred feet or less above the sea.* The specimens figured were gathered at Cap Martin in the end of October, 1863.

EXPLANATION OF PLATE XXXIII.—Fig. 1, calyx and bracts, with two joints of the branch. Fig. 2, the exterior bract.

---

* Statice pubescens, DC., ranges along the shores of Provence, Alpes Maritimes, and Western Liguria (Woods, Tour. Fl. p. 306), and may be gathered on the rocks at Nice, and on the isle Sainte-Marguerite at Cannes.

PLATE XXXIV.

GLOBULARIA ALYPUM, Linn.

Natural Order GLOBULARIEÆ.

Section of genus Alypum having a shrubby growth.

GEN. CHAR.—*Calyx* 5-lobed. *Corolla* monopetalous, hypogynous. *Stamens* 4, inserted on the tube of the corolla. *Ovary* free, 1-celled; *cell* containing 1 pendulous ovule. *Style* 1; *stigma* bifid. *Fruit* on utricle. *Embryo* straight; *radicle* turned towards the hilum. Koch (D.), Syn. Fl. Germ. ii. 512.

SPEC. CHAR.—*Receptacle* hairy, covered with linear, hairy, caducous scales. *Corolla* bilabiate; *upper lip* very short, bifid; *lower lip* long, ligulate, tridentate. *Growth* shrubby, perennial, 1–2 feet high.

Globularia Alypum, Linn. Sp. Plant. p. 139; Gren. et Godr. Fl. de Fr. ii. 756; Woods, Tour. Fl. p. 303.

HABITAT.—Warm rocky exposures on nummulitic and Jurassic limestone. October to March.

REMARKS.—Some authors—Grenier and Godron, Koch, and others —assert that Globularia has properly five stamens, the uppermost being reduced to a rudiment by abortion, and place Globularieæ close to Plumbagineæ. Lindley, however (Veg. Kingd. Lond. 1846, p. 666), considers the Order as more nearly related to the Teazels (Dipsaceæ). Maurice Wilkomm, in a monograph of Globularieæ, says, in a foot-note on page 11,—"According to the opinion of many authors, the stamens of Globularias are reduced to four, by the abortion of the upper stamen, which ought to be inserted between the segments of the upper lip; but with the exception of the rather prominent nerve in the upper lip of G. incanescens, which one might consider as an aborted stamen, one does not find in any species the least trace of a fifth sta-men." Globularia Alypum, Linn., is a rare European plant, found at intervals along the French shore of the Mediterranean, the western part of Italy, and in Sicily. I have also seen specimens from Greece and Smyrna in the Hookerian Herbarium at Kew.* The specimens figured were gathered above Pont St. Louis in November, 1864.

EXPLANATION OF PLATE XXXIV.—Fig. 1, the receptacle divided longitudinally, with one flower left adherent.

---

* Walpers (Ann. iii. 275) says that this species grows in northern Africa and western Asia, as well as in southern Europe.

PLATE XXXV.

CYTINUS Hypocistis, Linn.

Natural Order CYTINEÆ.

GEN. CHAR.—*Flowers* monœcious, bractate. *Perianth* campanulate; *æstivation* imbricate. *Stamens* adnate to a central column, which is united with the divisions of the perianth. *Ovary* inferior, 1-celled, 8 multiovulate parietal placentas.

SPEC. CHAR.—*Flowers* sessile, forming a capitulum. *Bracts* imbricate, fleshy below, frequently scorched at the tips, two at the base of each flower. *Growth* parasitic on Cistus.

Cytinus Hypocistis, Linn. Sp. Plant. p. 566; Gren. et Godr. Fl. de Fr. iii. 71; Woods, Tour. Fl. p. 324.

HABITAT.—Dry ground under Pine-trees. May.

REMARKS.—The relation in which Cytineæ stand to other plants has been very much canvassed, as, though their habit is that of Orobanche, Phelypea, etc., the structure of the flowers is essentially different. No sooner had Lindley constructed his class Rhizogens, which he considered as intermediate between Cryptogamous and Endogenous plants, and in which he placed Cytineæ between Balanophoreæ and Rafflesiaceæ, than he was met by most decided opposition in the very highest quarters. Both Robert Brown and Griffiths wrote at length (Trans. of Linn. Soc. xix.): the former placed Cytinus in a section of the Natural Order Rafflesiaceæ, along with the wonderful Rafflesia Arnoldi, and remarked (p. 229) that "the section Cytineæ seems to be unquestionably related to Asarineæ;" the latter declared himself adverse to the adoption of the class Rhizantheæ, in which he says, "a remarkable diversity of character has been sacrificed to an appearance resulting from the parasitism on roots and to an assumed absence of any ordinary form of vegetable embryo." At the time when these discussions took place, scarcely anything was known of the structure of the seeds of Cytinus, and I much regret that I have never been able to procure ripe fruits myself. I must therefore refer to Grenier and Godron (Fl. de Fr. p. 70) for the following information:—"Fruit sec ou bacciforme, uniloculaire, pulpeux intérieurement. Plusieurs graines logées dans la pulpe à teste dur coriace charnu. Embryon droit." I have followed Grenier and Godron (who coincide with Robert Brown) in placing Cytineæ among Dicotyledonous plants, believing their affinity to be nearer to Asarum and Aristolochia than to Fungus or Marchantia. It has been said that Cytinus resembles Fungi in the manner of its decay, but this is not strictly true, as the remains of its scales and stem may be found undecayed after the lapse of an entire year. There are two species belonging to the genus Cytinus mentioned by R. Brown, one from southern Africa and the other from equinoctial America. Cytinus Hypocistis is found on the roots of several species of shrubby Cistus in northern Africa, the Balearic Isles, Spain, South France, Italy, Dalmatia, Greece, and Turkey. It is said by Nyman (Syll. Fl. Eur.) to grow as far north as the Tyrol. Herbert mentions (Bot. Reg. xxxi. 85) having gathered it "on the rocky summit of Santa Decca, in Corfu." The specimen of Cytinus Hypocistis, Linn., and Cistus salvifolius, Linn., (Sp. Plant. p. 738), on which it is represented growing, was gathered in the Western Bay in May, 1863.

EXPLANATION OF PLATE XXXV.—Fig. 1, male flower and bracts. Fig. 2, female flower and bracts. Fig. 3, anthers.

Vincent Brooks

PLATE XXXVI.

FRITILLARIA involucrata, All.

Natural Order LILIACEÆ.

GEN. CHAR.—*Perianth* of six deciduous divisions, having at their bases nectariferous depressions. *Style* 3-cleft. *Seeds* flat.

SPEC. CHAR.—*Divisions of perianth* oblong. *Authers* apiculate. *Leaves* linear-lanceolate, opposite below, forming a whorl of three above.

Fritillaria involucrata, All. Auct. p. 34; Gren. et Godr. Fl. de Fr. iii. 180.

HABITAT.—Malaciers Mountain and Col di Tenda. May.

REMARKS.—The species of which the genus Fritillaria is composed seem to be remarkably subject to variation. Fritillaria montana, Hopp., is most closely related to the present plant, but it is believed to be distinguished by its only having leaves in pairs, being without the whorls from which this species is named.* As far as I can learn, F. involucrata is peculiar to the Maritime Alps.

EXPLANATION OF PLATE XXXVI.—Fig. 1, two divisions of the perianth of the natural size. Fig. 2, a stamen, magnified. Fig. 3, the style, magnified.

---

* F. messanensis, Rafin. ('Précis des Découvertes,' p. 44), is also very near to F. involucrata; and there are dried specimens at Kew, some of which have the whorl of three leaves above, but the lower leaves are alternate.

Plate XXXVII.

SCILLA ITALICA, Linn.

Natural Order LILIACEÆ.

GEN. CHAR.—*Divisions of perianth* free, deciduous, generally purple or blue. *Stamens* inserted on the base of the divisions of the perianth. *Pedicels* not jointed.

SPEC. CHAR.—*Perianth* spreading. *Bracts* in pairs underneath each flower, one of them being generally longer than the pedicel. *Leaves* 2–4.

Scilla italica, Linn. Sp. Plant. p. 442; Gren. et Godr. Fl. de Fr. iii. 186; Woods, Tour. Fl. p. 366.

HABITAT.—Rocky ground on the Tête du Chien, etc. Rarely found on the shore-level, but rather at from 1000 to 4000 feet. April.

REMARKS.—Scilla italica is generally very constant in colour, but last spring some fine plants were discovered at Mentone bearing pure white flowers, free from any tinge of purple.

In rocky places at Mentone we have Scilla maritima, Linn., a very fine plant, with a bulb as big as a large turnip and handsome spreading leaves. This plant is the officinal squill, the crystals which abound in the bulb having strong medicinal properties. Scilla italica, Linn., is found in southern France at Grasse and Frejus; in northern Italy; in Switzerland at Berne; and in Baden.

The specimens figured were gathered on the Mulaciers Mountain, in April, 1864.

EXPLANATION OF PLATE XXXVII.—Fig. 1, bracts and pedicel.

JM

Vincent brooks

PLATE XXXVIII.

ORNITHOGALUM NUTANS, Linn.

Natural Order LILIACEÆ.

Section of genus Ornithogalum having flowers in racemes.

GEN. CHAR.—*Divisions of perianth* persistent, white or greenish. *Stamens* inserted on the receptacle, slightly adherent to the bases of the divisions of the perianth. *Pedicels* not jointed.

SPEC. CHAR.—*Flowers* unilateral, nodding; *bracts* lanceolate, three times as long as the pedicels. *Stamens* 3 short, 3 long; *filaments* emarginate, petaloid, with the anther placed in the notch. *Capsule* broadly oval, 6-ribbed. *Leaves* linear, channelled.

Ornithogalum nutans, Linn. Sp. Plant. p. 441; Gren. et Godr. Fl. de Fr. iii. 189; Woods, Tour. Fl. p. 360.

HABITAT.—Meadows in Sospello Valley. April.

REMARKS.—Ornithogalum nutans, Linn., is a plant of very wide range, and appears to have considerable power of naturalizing itself in countries into which it is introduced. It is perhaps to this faculty that it owes its place in English and some other floras. It is much spread over central Europe, from the Baltic to the Mediterranean, growing in Gottland, Denmark, Belgium, Germany, Austria, and Hungary. It reaches also from Castile in Spain, through France, Italy, Croatia, Hungary, Transylvania, Greece, and Asia Minor, to the south of Russia (Nyman, Syll. Fl. Eur. p. 371). There are dried specimens in the Hookerian Herbarium at Kew, gathered among the crops at Smyrna, at Arsa, in southern Anatolia, communicated by Professor Forbes, and one extracted from the Aucher-Eloy Herbier d'Orient, found at Moglah.

EXPLANATION OF PLATE XXXVIII.—Fig. 1, one of the longer stamens.

PLATE XXXIX.

ALLIUM NEAPOLITANUM, Cyrill.

Natural Order LILIACEÆ.

GEN. CHAR.—*Divisions of perianth* 6, spreading or bell-shaped. *Stamens* inserted on or close to the base of the divisions of the perianth, *filaments* more or less dilated, the 3 innermost sometimes 3-fid ; *anthers* attached by the back. *Flowers* in umbels. *Spathe* of one or more valves.

SPEC. CHAR.—*Umbel* dense. *Spathe* 1-valved. *Divisions of perianth* spreading, ovate, very obtuse. *Leaves* broadly linear, acuminate ; margins finely denticulate. *Stem* smooth, triangular, 2 of the angles acute and 1 obtuse. *Bulb* sometimes enclosing bulblets in its membranous covering. Gren. et Godr. Fl. de Fr. iii. 205.

Allium neapolitanum, Cyrill. Pl. Rar. Neap. Fasc. i. p. 13. t. 4 (1788) ; Gren. et Godr. Fl. de Fr. iii. 205 ; Woods, Tour. Fl. p. 371.

HABITAT.—Near the shore, under Lemon- and Orange-trees. March, April.

REMARKS.—This species is exempt from the strong garlic smell so generally possessed by members of this genus, but the juicy leaves and stem have a peculiar and disagreeable odour when bruised or held long in the hand. Allium neapolitanum, Cyrill., ranges from Andalusia and Granada, along the northern coast of the Mediterranean, to Greece, the Archipelago, and Palestine. It has been found on the banks of the Guadalquivir,[*] at Toulon, Hyères, Cannes, Narbonne, Corsica (Gren. et Godr. Fl. de Fr. iii. 205), near the aqueduct at Genoa, at Rome, Naples, Trau and Spalatro in Dalmatia (De Visiani, quoted in Rchb. Ic. Fl. Germ.), in Chios (Jaub. and Spach, Ill. Plant. Or.), and at Jerusalem.[†] The specimen figured was gathered near Pont St. Louis, early in April, 1864.

EXPLANATION OF PLATE XXXIX.—Fig. 1, a seed of the natural size. Fig. 2, the same, magnified. Fig. 3, margin of leaf, magnified.

---

[*] There are specimens in the Hookerian Herbarium at Kew, gathered there by Bourgeau.

[†] In the Kew herbarium there is a specimen from Jerusalem of Boissier's collecting.

1.   *2*

PLATE XL.

CROCUS VERSICOLOR, Gawl.

Natural Order IRIDACEÆ.

GEN. CHAR.—*Perianth* regular, campanulate, with a long tube. *Stamens* 3. *Style* filiform, springing nearly from the bulb. *Stigma* 3-lobed, generally more or less cut.

SPEC. CHAR.—*Flower* springing from a spathe of two scarious sheathing scales, of which the outer one is obtuse and larger, the inner narrow and acute. *Perianth* smooth at the throat. *Anthers* linear, sagittate, rather longer than the smooth filaments. *Stigmas* entire or slightly cut. *Leaves* appearing with the flowers. *Bulb* covered with fibrous tunics ("Fibres of coating nearly parallel," Woods, Tour. Fl. p. 357). Gren. et Grodr. Fl. de Fr. iii. 257.

Crocus versicolor, Gawl. Bot. Mag. 1110 ; Gren. et Godr. Fl. de Fr. iii. 257 ; Woods, Tour. Fl. p. 357.

HABITAT.—Rocky ground close to the sea on Cap Martin ; abundant also, at an elevation of from 3000 to 4000 feet, on most of the mountains. February.

REMARKS.—The colour and even in some measure the growth of this plant is given to vary, the flowers assuming yellowish or purple tints, or appearing of a pure white, while sometimes two flowers spring from one bulb. Crocus versicolor, Gawl., and C. medius, Balb., figured in Part I., are the only representatives of the genus which have as yet been found in the immediate neighbourhood of Mentone. There are specimens of C. versicolor in the herbarium at Kew from Dalmatia and Tiflis, south of the Caucasus ; and Grenier and Godron cite it from Draguignan, Grasse, Frejus, Toulon, and Aix.

EXPLANATION OF PLATE XL.—Fig. 1, stigma and part of style. Fig. 2, fibrous coating of the bulb.

PLATE XLI.

NARCISSUS INTERMEDIUS, Lois.

Natural Order AMARYLLIDACEÆ.

GEN. CHAR.—*Perianth* of 6 spreading divisions, within which is a bell-shaped crown. *Stamens* 6, inserted either on the crown or base of the tube of the corolla.

SPEC. CHAR.—" Flowers 2–5; *crown* cup-shaped, margin wavy, about one-third of the divisions of the perianth. *Leaves* green, channelled, nearly semicylindrical on the back " (Lois. Fl. Gall. i. 236).

Narcissus intermedius, Lois. Fl. Gall. i. 236 t. 6; Gren. et Godr. Fl. de Fr. iii. 258; Woods, Tour. Fl. p. 361.

HABITAT.—Shady, near water in Eastern Bay. March.

REMARKS.—This is an extremely rare Mentonese plant, and is one which hitherto has been supposed to belong exclusively to western France. Only one habitat has as yet been discovered here, but I hope that by the aid of this drawing, others may possibly be found. The bank on which it grows is covered with reeds (Arundo Donax, Linn.), and quite free from cultivation, so that I regard it as a truly wild species. The specimens I have seen never quite coincided with the description of Loiseleur's N. intermedius, quoted above, but they agree more nearly with this species than any other known to me. The leaves were perfectly cylindrical in their whole length, slightly ribbed, and the cellular tissue spongy with lacunæ at regular intervals. Many botanists believe that there exists a complete series of forms intermediate between Narcissus Jonquilla, Linn. Sp. Plant. p. 417. Narcissus intermedius is said only to grow on the hills near Bayonne and the landes of Dax.

EXPLANATION OF PLATE XLI.—Fig. 1, stigma and part of style. Fig. 2, a cross section of the leaf. Both the figures are magnified.

PLATE XLII.

ORCHIS PROVINCIALIS, Balb.

Natural Order ORCHIDACEÆ.

Section of genus Orchis having separate glands.

GEN. CHAR.—*Perianth* ringent, hooded. *Lip* 3-lobed, spurred. *Glands* of the stalks of the pollen-masses in a common pouch. Bab. Man. of Brit. Bot. p. 306.

SPEC. CHAR.—*Flowers* pale-yellow. *Spike* lax. *Petals* and *sepals* reflexed or expanded. *Lip* deeply 3-lobed, dotted with purple points, the terminal lobe emarginate, generally with a tooth in the notch; *spur* nearly horizontal, about equal to the ovary. *Leaves* linear-lanceolate, generally spotted.

Orchis provincialis, Misc. Alt. Taur. p. 20; Gren. et Godr. Fl. de Fr. iii. 293; Woods, Tour. Fl. p. 351.

HABITAT.—Among grass on the eastern side of Montegrosso, etc. April.

REMARKS.—Orchis provincialis, Balb., differs but little from Orchis pallens, Linn. (Mant. p. 292), the latter having broader leaves, free from spots, and a less deeply divided lip, without purple points. Orchis pallens, Linn., is not found nearer to Mentone than the Sospello Valley, where it enjoys a cooler and moister air. Orchis provincialis, Balb., ranges from the westernmost to the north-easternmost shores of the Mediterranean, growing in Spain, Southern France, Italy, Dalmatia, Greece, Crete, on Mount Athos, and at Constantinople (Nyman, Syll. Fl. Eur. 1845, p. 357). I have seen specimens at Kew from Rome, presented by Mr. Woods; from Monte Pizzuta, above Piana (Sicily), collected by Huet du Pavillon; and one gathered in Attica by Boissier.

EXPLANATION OF PLATE XLII.—Fig. 1, an entire flower and bract, magnified.

Vincent Brooks, Imp.

PLATE XLIII.

OPHRYS INSECTIFERA, Linn.

Natural Order ORCHIDACEÆ.

GEN. CHAR.—*Lip* without a spur. *Pollen-masses* having 2 separate glands enclosed in 2 apparently distinct pouches. *Ovary* not twisted.

SPEC. CHAR.—*Sepals* spreading. *Lip* entire or lobed. *Column* prolonged beyond the anther-cells into a beak of variable length.

Ophrys insectifera, Linn. Sp. Plant. ii. 1343 (1765).

HABITAT.—Banks in western and eastern bays, from shore-level to about 1000 feet. December to March.

REMARKS.—I have devoted this Plate and the two following to drawings illustrative of a few of the forms which Linnæus considered as varieties sprung from one original type, but which have since his time been always arranged as a greater or less number of species. Of these the principal with which the present inquiry has to do are Ophrys aranifera, Huds. (Fl. Angl. ed. 2 (1778), p. 391), the Spider Ophrys; O. arachnites, Reichard (Fl. Mœnofrancof. ii. 89); O. scolopax, Cav. (Ic. ii. p. 46, t. 161), and O. apifera, Huds., the Bee Ophrys. Linnæus included in this manner, under the same name, plants which, as he said, "seem at the first glance perfectly distinct; but," he continues, "one who compares them with their congeners, and has before him all the varieties at the same moment, will easily perceive them to be sprung from one stock, and will find no means by which he may distinguish them, however constant they (the varieties) may be" (Linn. Sp. Plant. ii. 1344). I have never had any opportunity of studying the degree of variation to which Ophrys muscifera, Huds., may be subject, and I wish it clearly to be understood that my present object is confined to an attempt to show something of the intermediate forms between O. aranifera, Huds., and O. apifera, Huds., without attempting any consideration of adjacent varieties. It is scarcely necessary to remark that in three plates it is impossible to do more than supply a few examples of the manner of variation, and I have therefore selected such forms as have not previously been figured, so as to add to the information already supplied on this subject by Reichenbach, in his 'Icones Floræ Germanicæ,' vols. xiii. xiv., by the 'Botanical Magazine' and 'Register,' and by Sowerby's 'English Botany.' There are three organs in the flower of O. insectifera, Linn., on which specific characters have chiefly been founded by modern botanists: 1, the petals, 2, the lip; 3, the anther. In the petals the glabrous and flat form changes till it becomes pubescent and recurved, and the linear-oblong shape is modified into ovate or cordate. The lip, even in the forms which flower earliest and which best represent O. aranifera, Huds., is very variable, being either bossed or not, lobed or entire ; the process may also be found in every stage of growth, advancing with the season (for this see Remarks on Plate XLIV.), from the tooth-like point of the January plant to the tridentate or entire lobe of those flowering in May. The markings also undergo a gradual modification, so that the lines resembling the Greek letter π, which are found in all the plants figured in this Plate, vary by the approximation of the parallel limbs until the complex figuring of O. apifera, Huds., is obtained. At one time I thought that O. apifera, Huds., might be separated from its congeners by the curvature of the terminal lobe of the lip, which is generally so recurved that the process is hidden beneath ; but I found specimens in which this character was scarcely evident, and I discovered in Reichenbach's figures (Ic. Fl. Germ. xiii. 96) a plant which he calls O. apifera, var. Trollii, in which the entire lip is porrect or very slightly curved, as in O. scolopax, Cav. The anther also is a variable feature, and cannot be depended on for characters to separate O. apifera, Huds., from the forms nearest to it. It is true that the column is generally prolonged into a beak, but this is very inconstant ; and it is also true that the pollen-masses are spontaneously released from the cells, but this takes place also at times in O. scolopax, Cav. (Cont. to Fl. of Mentone, i. 19). (Continued in " Remarks" on the two following Plates.)

EXPLANATION OF PLATE XLIII.—Fig. A 1, the entire lip of form A. Fig. A 2, anther and stigmatic chamber of the same. Fig. A 3, petal of the same. Fig. B 1, lobed lip of form B. Fig. D 1, incurved lip of form D. All the figures are magnified.

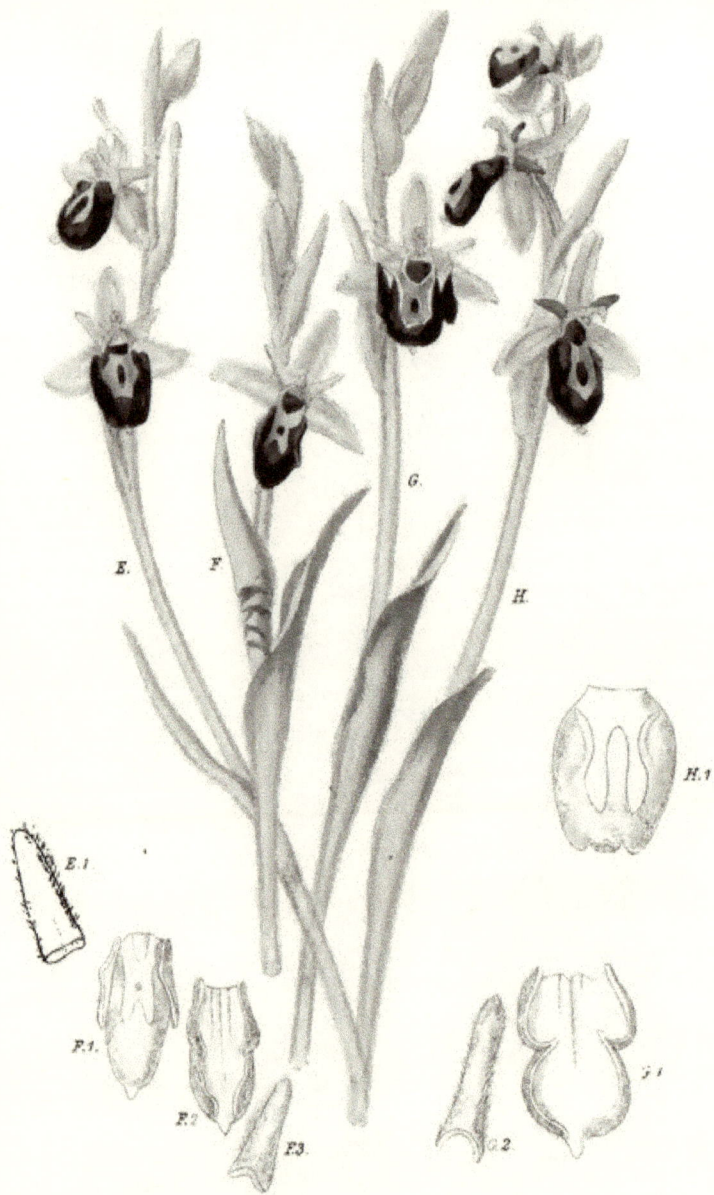

E.   F.   G.   H.

E.1   F.1   F.2   F.3   G.2   G.1   H.1

PLATE XLIV.

OPHRYS INSECTIFERA, Linn.

Natural Order ORCHIDACEÆ.

GEN. CHAR.—See description of Plate XLIII.

SPEC. CHAR.—Ditto.

HABITAT.—Banks in western and eastern bays, from shore-level to about 600 feet elevation. End of March and April.

REMARKS.—During the past spring (1865) I watched almost from day to day the development of flower after flower, and carefully noted the dates at which the varying forms appeared. I had known from my previous three years' experience that O. aranifera, Huds. (represented at letter A, Plate XLIII.), comes into flower in December, and O. apifera, Huds. (represented at letters M and N in Plate XLV.), in May, while the intermediate varieties appear in February, March, and April respectively; but it was not till the present season that I became thoroughly convinced of the intimate connection which exists between the sequence in which each variety flowers and its approximation to one of the extreme forms. Thus at Mentone a definite order seems to be maintained among these forms, and one which accords with their respective degrees of similarity to form A. The rule is that during January and February there is but little change, except that towards the end of the latter month the sepals grow pale and lose the green colouring-matter, and the petals become purplish-brown. Early in March, flowers with whitish or pink-tinted petals and sepals appear, and by the end of that month the petals, which were flat and smooth before, become slightly pubescent and occasionally a trifle recurved. As soon as April is well commenced, the lip, which has already shown more or less rudimentary stages of a terminal process, becomes strongly apiculated and variously bossed and lobed; the petals become reflexed, downy, and even velvety, and the varieties creep, step by step, onwards towards O. scolopax, Cav. (represented at letter K, Plate XLV.), which appears quite at the end of the month, and forms in its varying characters the nearest link to O. apifera, Huds., the latest of all. I cannot regard this sequence as fortuitous, but rather am led to surmise that laws new to me are here in operation. (Continued in "Remarks" on the following Plate.)

EXPLANATION OF PLATE XLIV.—Fig. E 1, the flat, ciliated petal of form E. Fig. F 1, lip of form F, seen from above. Fig. F 2, the same, seen from below. Fig. F 3, the pubescent, slightly curved petal of form F.   Fig. H 1, the lip of form H, showing the upturned process.   Fig. G 1, the lobed and bossed lip of form G, seen from below. Fig. G 2, the strongly recurved, pubescent petal of form G.   All the figures are magnified.

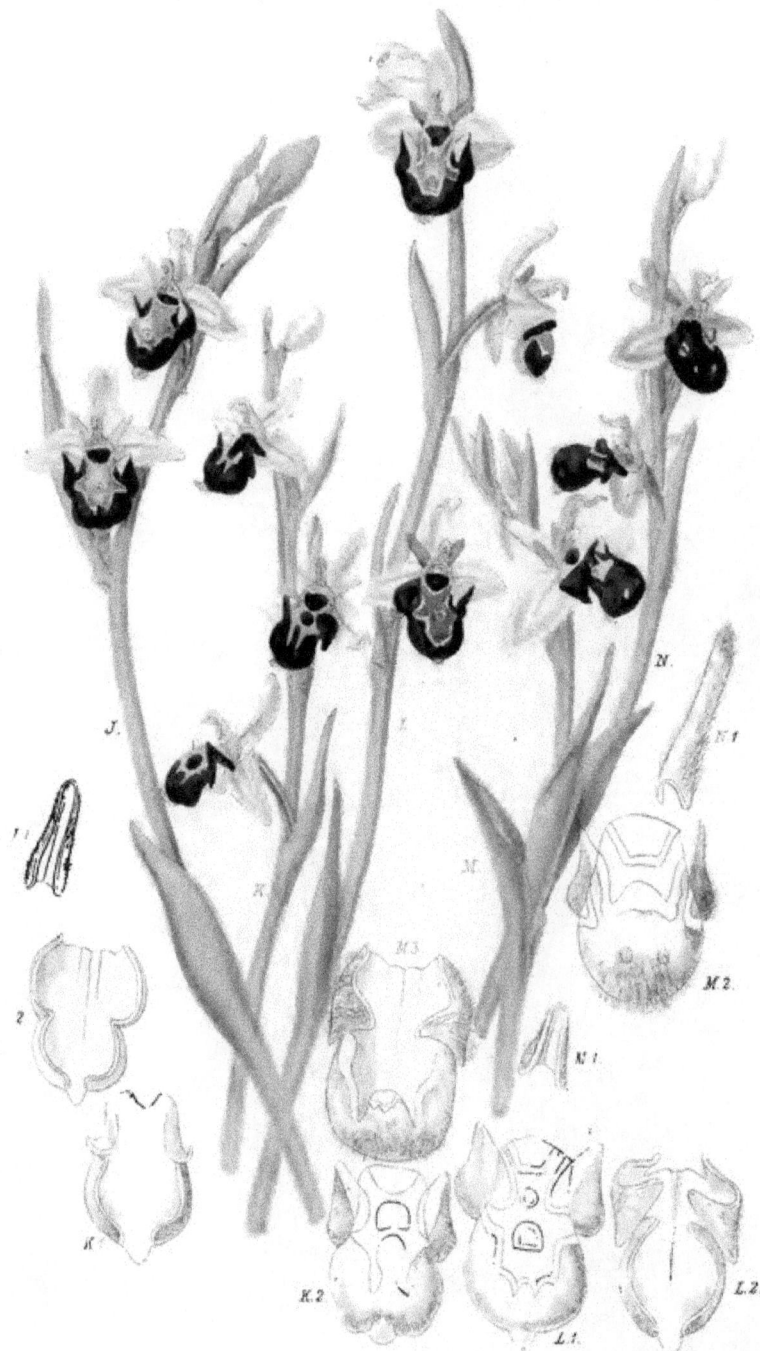

PLATE XLV.

OPHRYS INSECTIFERA, Linn.

--------- — ----

Natural Order ORCHIDACEÆ.

GEN. CHAR.—See description of Plate XLIII.
SPEC. CHAR.—Ditto.
HABITAT.—Banks in western bays, from shore-level to about 600 feet
elevation. End of April and May.

REMARKS.—In this plate two specimens of Ophrys apifera, Huds.,
are figured at M, N. The spike figured at N was drawn from one of
seven spikes gathered at Reigate on my return to England, four of
which had the long petals figured at N 1 ; the other specimen was from
Mentone, and had the almost cordate petal, which is so generally
thought to distinguish the Bee Ophrys. As all the forms figured seem
tolerably abundant and distinct, I shall attempt an arrangement of the
characters by which they may frequently be distinguished. I also
append their respective dates of first expansion and proximate synonyms.

Sect. 1. *Petals* flat, smooth.
a. *Petals* green. *Lip* entire or slightly emarginate. (End of December.) O. aranifera, Huds.; O. aranifera, a. genuina, Rchb. Ic. Fl.
   Germ. xiii. 88.
b. *Petals* brownish. *Lip* deeply lobed and bossed. (End of January.)
c. *Sepals* tinged with colour. *Lip* entire, apiculate. (March 8th.)
   O. atrata, Lindl. Bot. Reg. 1087, *partly.*
d. *Lip* elongate, margins inflexed. (March 23rd.)

Sect. 2. *Petals* puberulent or ciliate.
e. *Petals* ciliate, flat. (March 23rd.)

Sect. 3. *Petals* recurved, pubescent. *Terminal lobe of lip* porrect.
f. *Lip* lobed, bossed, elongate ; *lobes* adpressed. (April 1st.)
g. *Lip* lobed ; *bosses* conical. (April 1st.)
h. *Lip* entire, margins slightly recurved ; *apiculum* upturned. (April
   9th.) O. arachnites, Reichard (Fl. Mœnofrancof. ii. 89) ; O. fuci-
   flora, Rchb. (Ic. Fl. Germ. xiii. p. 85. t. 461), *partly.*
j. *Petal* strongly recurved. (May 5th.)
k. *Process* of lip tridentate. (May 1st.) O. scolopax, Cav. (Ic. ii. p.
   46. t. 161).
l. *Bosses* free from the terminal lobe of the lip. (May 5th.)

Sect. 4. *Terminal lobe of lip* bent underneath.

m. Petal not half as long as the anther. *Process of lip* slightly tridentate. (May 6th.) O. apifera, Huds., *partly.*

n. Petal about equal to anther. (June 22nd.)

It is supposed by some botanists that many of the forms under which these most variable plants appear are only hybrids between a greater or less number of species ; others, again, believe that though an occasional cross is probably a most efficient stimulus to variation, the greater part, if not all, of these individuals are the product of natural selection.

EXPLANATION OF PLATE XLV.—Fig. J 1, petal of form J. Fig. J 2, under side of lip of form J. Fig. K 1, under side of lip of form K. Fig. K 2, upper side of lip of form K. Fig. L 1, upper side of lip of form L. Fig. L 2, under side of the same. Fig. M 1, petal of form M. Fig. M 2, upper side of lip of form M. Fig. M 3, under side of the same. Fig. N 1, petal of form N.

PLATE XLVI.

OPHRYS FUSCA, Link, and OPHRYS LUTEA, Cav.

Natural Order ORCHIDACEÆ.

GEN. CHAR.—See description of Plate XLIII.

SPEC. CHAR.—*Flowers* few. *Sepals* green, the uppermost forming a hood over the anther. *Petals* linear, yellow, green or brown. *Lip* wedge-shaped, nearly flat, lobed, gibbous at the base, velvety except on the bilobed marking and the narrow margin. *Anther* without any beak.

Ophrys fusca, Link in Schrad. Diar. p. 2 (1799) ; Gren. et Godr. Fl. de Fr. iii. 305 ; Woods, Tour. Fl. p. 353.

SPEC. CHAR.—*Petals* yellow, linear-oblong. *Lip* nearly as broad as long, margin wide, smooth, wavy. *Anther* without any beak.

Ophrys lutea, Cav. Ic. ii. p. 46, t. 160 ; Gren. et Godr. Fl. de Fr. iii. 305 ; Woods, Tour. Fl. p. 353.

HABITAT.—Sunny banks near the shore. March.

REMARKS.—The two plants figured on this Plate were included by Linnæus under Ophrys insectifera ; and, different as they seem, I have some reason to suppose that a more accurate knowledge of the intermediates may at some future day lead me to enter into and understand this view. For the present, however, I treat O. lutea, Cav., (the plant figured on the right-hand,) and O. fusca, Link, (that on the left,) as distinct from one another and from all others, though, as this Plate shows, O. fusca, Link, varies considerably in the direction of O. lutea, Cav. Probably O. muscifera, Huds., is a near approach to O. fusca, Link, but this plant has not been found at Mentone, and I have never had the opportunity of studying it in life. O. fusca, Link, and O. lutea, Cav., are found in Lusitania, southern and south-western France, Italy, Sardinia, Sicily, Dalmatia, Greece, and Crete.

EXPLANATION OF PLATE XLVI.—Fig. 1, lip of O. fusca, Link. Fig. 2, anther and stigmatic chamber of the same. Both figures magnified.

PLATE XLVII.

ARUM ARISARUM, Linn.

Natural Order AROIDEÆ.

Section of genus Arum having a tubular spathe, 1-celled anthers, and a stigma borne on a distinct style.

GEN. CHAR.—*Spathe* hooded. *Spadix* naked above. *Perianth* wanting, male flowers consisting of a stamen only and the females of a pistil. *Stamens* in several rows, placed above. *Pistils* placed below at the base of the spadix. *Berry* 1- or many-seeded.

SPEC. CHAR.—*Spathe* nearly cylindrical, bent above, borne on a peduncle often as long as the leaves. *Spadix* scarcely shorter than the spathe, bent above. *Berries* few, angular. *Leaves* cordate or hastate at the base; *auricles* obtuse.

Arum arisarum, Linn. Sp. p. 1370; Gren. et Godr. Fl. de Fr. iii. 331; Woods, Tour. Fl. p. 378.

HABITAT.—From shore-level to between 3000 and 4000 feet elevation. Autumn and winter.

REMARKS.—Almost every traveller along the Riviera notices the strange appearance of this plant, and not a few bestow upon it some fanciful name of their own. At Mentone the country-people call the flowers Capucini, in quaint allusion to the brown-hooded *Frati* of the neighbouring town of Ventimiglia. We have at Mentone but one other species of Arum, A. italicum, Mill. (Dict. n: 2), a plant which closely resembles our own A. maculatum, Linn. (Sp. Plant. p. 1370). A. arisarum, Linn., grows in Granada, Lusitania, the Balearic Isles, South France, Italy, Dalmatia, and Greece (Nyman, Syll. Fl. Eur.).

EXPLANATION OF PLATE XLVII.—Fig. 2, the lower part of the spadix, bearing the flowers, the spathe being cut away.

PLATE XLVIII.

ANDROPOGON Allionii, DC.

— —

Natural Order GRAMINEÆ.

Section of genus Andropogon having solitary or fasciculate spikes at the summit of the culm.

GEN. CHAR.—" *Spikelets* in pairs or in threes above, of one flower each, polygamous, the female and hermaphrodite spikelets always sessile. *Outer glumes* nearly equal, either awnless or the uppermost awned. *Flowering glumes* membranous, the lower one rounded on the back, the upper shorter and awned in the fertile flowers. *Paleæ* 2, smooth." Gren. et. Godr. Fl. de Fr. iii. 465.

SPEC. CHAR.—*Spike* solitary, terminal. *Spikelets* in pairs, lower ones all male, upper ones male and female, the former stalked and the latter sessile. *Female spikelet* bearded at the base; *outer glumes* stiff, nearly equal, brown, rough with hairs, the uppermost membranous at the edge; *flowering glumes* much shorter, the uppermost awned, entire; *awns* as long as the spike, twisted, slightly attached to the interior of the glume at the base.

Andropogon Allionii, DC. Fl. Fr. iii. 97; Gren. et Godr. Fl. de Fr. iii. 467; Woods, Tour. Fl. p. 395.

HABITAT.—Jurassic limestone rocks on the eastern side of Pont St. Louis. October, November.

REMARKS.—The genus Andropogon is said to have representatives in every quarter of the globe, inhabiting the subtropical regions. Alph. de Candolle mentions a member of this genus in America, and says that Kunth records the same plant from Africa (Geog. Bot. ii. 1046), while Dr. Hooker describes moors and turfy spots in India entirely covered with certain species (Himalayan Journ.). The present species is a decidedly rare European plant, growing on warm rocks near Susa at the foot of Mont Cenis, in the Tyrol, on the shores of Lake Lugano (Parlatore, Fl. It. i. 149), thus reaching as far north as the Canton Tessin in Switzerland, whence it is cited by Nyman, who also records it in continental Italy, Sicily, and Dalmatia (Syll. Fl. Eur. 404). The specimens figured were gathered near Pont St. Louis, in November, 1863.

EXPLANATION OF PLATE XLVIII.—Fig. 1, two male spikelets, from the lower part of the spike. Fig. 2, one stalked male and one sessile female spikelet. Fig. 3, pistil and flowering glumes, with part of the awn. Fig. 4, the awn, showing its slender attachment to the base of the uppermost flowering glume. Fig. 5, an immature grain, with the stigmas cut away. Fig. 6, the same, mature. All the figures magnified.

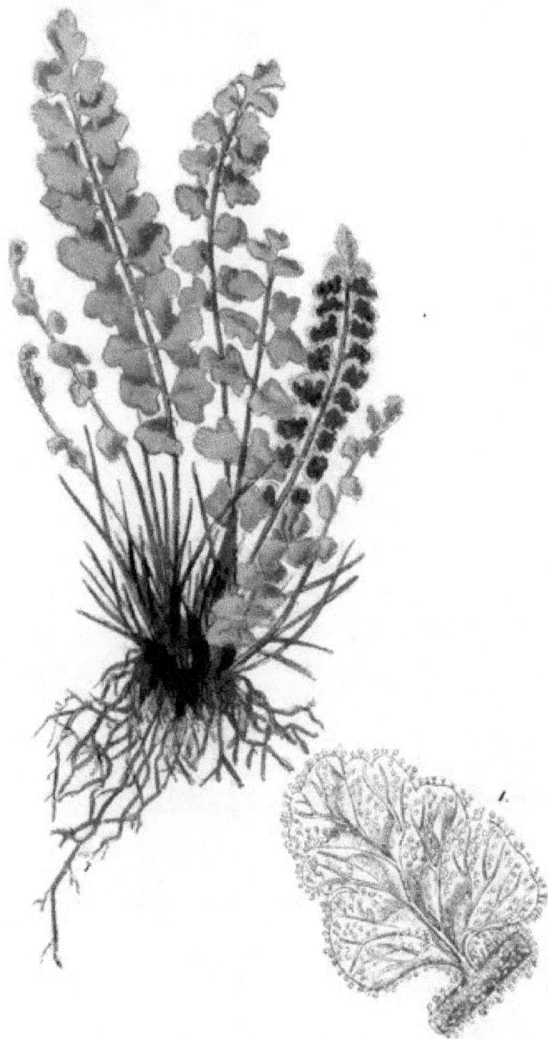

Plate XLIX.

ASPLENIUM Petrarchæ, DC.

Natural Order Filices.

Gen. Char.—*Sori* linear or oval, few and solitary, rarely regularly placed in two rows. *Indusium* fastened by the outer edge, free on the inner edge (*i. e.* towards the central nerve of the pinna), bent outwards. *Fronds* 1-2-3-pinnatifid.

Spec. Char.—*Fronds* rarely 3 inches long. *Rachis* black in the greater part of its length, covered with glandular hairs, cylindrical (not channelled and winged as the rachis in A. Trichomanes, Linn., is). *Indusia* frequently placed below the branching of the nerves.

Asplenium Petrarchæ, DC. Fl. Fr. v. 238. *A. Trichomanes*, β. *pubescens*, Gren. et Godr. Fl. de Fr. iii. 636.

Habitat.—Sunny rocks of Jurassic limestone, from shore to 1290 feet elevation. Winter.

Remarks.—MM. Grenier and Godron treat this plant as a mere variety of A. Trichomanes, Linn. (Sp. Plant. p. 1540), but though both these ferns grow together at Mentone, I have been quite unable to trace any intermediates, and after careful examination suggest the above characters as distinctive. A. Petrarchæ, DC., seems to be a rare plant, though perhaps it may be often overlooked. Grenier and Godron cite it as growing at Vancluse and Toulon.

Explanation of Plate XLIX.—Fig. 1, one of the pinnæ and part of the rachis, magnified.

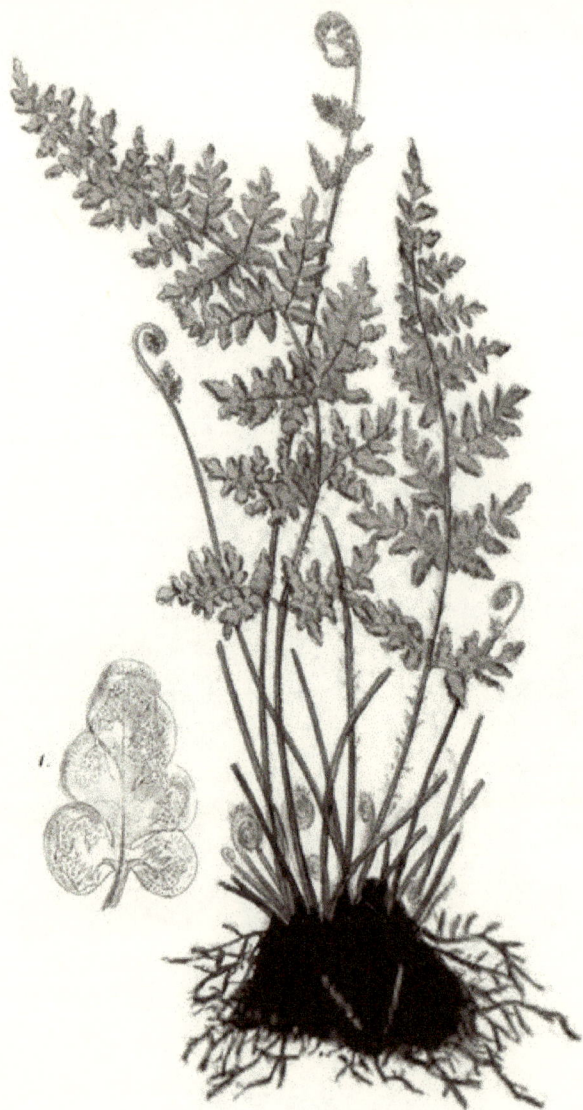

PLATE L.

CHEILANTHES ODORA, Sw.

Natural Order FILICES.

GEN. CHAR.—" *Segments* of the fertile fronds recurved and membranous at the edges, partly covering the sori. *Sori* forming a line, which follows the edges of the segments and leaves; the centre of the limb bare.' Gren. et Godr. Fl. de Fr. iii. 641.

SPEC. CHAR.—*Frond* 4-7 inches long, very smooth, tripinnate below; *segments* oval, petiolate, coriaceous. *Rachis* sparsely covered with scaly hairs.

Cheilanthes odora, Sw. l. c. pp. 127 and 317; Gren. et Godr. Fl. de Fr. iii. 641 ; Woods, Tour. Fl. p. 426.

HABITAT.—Terrace walls, from 100 to 1334 feet elevation. Winter and spring. ·

REMARKS.—Cheilanthes odora, Sw., seems to be quite a southern plant, though it advances as far as the Pyrenees, and even to Susa, at the foot of Mont Cenis. Nyman gives the following stations :—Spain, Italy, Dalmatia, Greece, and Mount Athos in Turkey.

EXPLANATION OF PLATE L.—Fig. 1, one of the segments, seen at the back.

PLATE LI.

ANEMONE CORONARIA, Linn.; var. β. *cyanea*, Ardoino.

Natural Order RANUNCULACEÆ.

GEN. CHAR.—See description of Plate I. Part I.

SPEC. CHAR.—*Flower* solitary, purplish-blue. *Leaves of involucre* finely divided, sessile. *Sepals* 5–7. *Leaves* finely divided into linear cuneate-lanceolate segments.

Anemone coronaria, β. *cyanea*, Ardoino. Fl. Alpes Mar. p. 12. *A. cyanea*, Risso, Fl. de Nice, p. 7. *A. coronarioides*, Hanry (ex Ardoino, l.c.).

HABITAT.—Castellar, near Mentone, where I gathered these specimens in January, 1865.

REMARKS.—The different varieties of Anemone coronaria, Linn., or Parsley-leaved Anemone, are but slightly characterized, yet several forms have been distinguished, and even raised to the rank of species. In certain localities particular forms grow to the exclusion of others, and constitute there marked colonies. The facility with which these plants are multiplied by division of their rootstocks tends to preserve the different races, some of which are, owing to their sterility, very dependent upon this mode of propagation. At Mentone I have never been able to procure good seed from any variety of A. coronaria, Linn. Along the Riviera both this plant and A. hortensis, β. fulgens, Gren. et Godr., are principally to be found in cultivated terraces and olive yards.

Though now generally admitted as a native of France and Italy, we are told by M. Alph. De Candolle,[*] that Anemone coronaria, Linn., has probably spread westward from Greece, Constantinople, and Asia Minor within recent times, as it was excluded from the lists of indigenous plants given by ancient authors. Near Bologna, however, Bertoloni (Fl. Ital. v. 456) describes the wild hillsides as being full of scarlet, purple, and white varieties of A. coronaria, Linn.!

On the Turbit Mountain, near Mentone, a few terraces are occupied by a variety having pale whitish flowers streaked with purplish-pink, and remarkable for having the bases of the sepals so much curved that the flower is quite cupped inwards below. The leaves also have a peculiar appearance, and this variety has been distinguished as a species (A. rosea, Hanry). I have received specimens every spring during the last six years from this locality, and the form certainly remains true.

* Alph. De Candolle, 'Géographie Botanique,' ii. 637.

A form with small double pink and greenish flowers is not uncommon, and is probably the result of cultivation upon A. rosea, Hanry. This is the var. γ. Rissoana of Ardoino. It would be a matter of some interest to collect and figure all the varieties and subvarieties which grow along the coast, as many have probably been overlooked. I have, for instance, observed a subvariety of the variety figured, which constantly reappears on the property of Carnoules, near Mentone, with flowers of a peculiar and very rich purple. There are also scarlet and purple-blue-flowered varieties, which may be recognized by their leaves alone, the segments being much broader than those of the form figured. In the fields of Mouans, near Cannes, several very fine varieties grow, some with double flowers being of especially gorgeous colours. It is a matter of wonder to me that these beautiful plants are not introduced into the public gardens of the health stations of the Riviera, where they would certainly give great pleasure to the visitors.

Till lately I believed that the double forms of Anemone were only to be found in cultivated ground; but I have since heard from the Rev. T. Butler that he has gathered double-flowered specimens of A. sulphurea, Linn. (now generally considered a mere variety of A. alpina, Linn.), between Saas and the Mte. Moro, in a thoroughly wild habitat. He says, "It grew just where the glacier crosses the valley and forms the Matmark See by damming up the stream" (a tributary of the Visp). The surrounding ground was alpine in its nature and "the staple commodities were Gentians, Viola calcarata, Anemone alpina and sulphurea, stones edged with Ranunculus alpestris and plants of that class."

Mr. Butler was kind enough to send me a specimen of this double Anemone for inspection. The flower consisted of five sepals of the normal broadly ovate shape, and within these of a hemispherical mass of linear-lanceolate segments, all very silky on the back, as were the sepals. In herbaria I have often seen double-flowered specimens of A. nemorosa, Linn., but always from botanic gardens.

Dr. Seemann (Journ. of Bot. ii. 1864, pp. 176–187 and 318) enumerates a large number of plants which present double flowers, but these probably are nearly all garden examples. However, plants of several kinds have been observed presenting double flowers in truly wild stations; for example, I have a specimen of Anagallis tenella, Linn., the flowers of which are double, gathered in co. Kerry by Dr. Ch. Battersby.

C

C.1.

B.

B.1.

B.2.

B.3.

A.

A.1.

PLATE LI. *bis.*

(A) ANEMONE palmata, Linn.; (B) A. trifolia, Linn.;
(C) A. ranunculoides, Linn.

Natural Order Ranunculaceæ.

Gen. Char.—See description of Plate I. Part I.

(A.) Spec. Char.—*Flowers* 1-2 from each scape (stem?); *leaves of involucra* sessile, irregularly 3-lobed; *scape* covered with silky yellow hairs above, when branched bearing a second involucre. *Sepals* about 10, oblong-lanceolate, subacute, outer surface silky. *Achenes* (immature) hairy, ending in a long glabrous linear style. *Leaves* appearing with the flowers, leathery, thick and waxy, shortly 3-lobed, reniform, usually broader than long. *Rootstock* thick, often bearing long tuber-like cylindrical branches, somewhat flattened at the ends.

Anemone palmata, Linn. Sp. Plant. p. 758; Gren. et Godr. Fl. de Fr. i. 14; Woods, Tour. Fl. p. 3.

(B.) Spec. Char.—*Flower* 1; *leaves of involucre* stalked, each of 3 ovate regularly-toothed leaflets. *Sepals* about 6, oblong obtuse, the three inner usually alternating with the three outer. *Anthers* nearly white. *Achenes* pubescent, ovate-elliptic, compressed, the glabrous style forming a short tapering point; at the base of the achene there is a short process at the point of attachment with the receptacle. *Leaves* appearing after the flowers, of 3 ovate leaflets, like those of the involucre. *Rootstock* horizontal, jointed, whitish, brittle.

Anemone trifolia, Linn. Sp. Plant. p. 762; Woods, Tour. Fl. p. 3.

(C.) Spec. Char.—*Flowers* 1-5; *leaves of involucre* oblong-lanceolate, deeply and irregularly toothed; no second involucre. *Sepals* 5-6, obovate obtuse, pubescent beneath. *Achenes* subglobular, ending in a tapering glabrous style. *Leaves* appearing after the flowers similarly shaped to those of involucre. *Rootstock* angularly branched (not given complete in figure from want of space).

Anemone ranunculoides, Linn. Sp. Plant. p. 762; Gren. et Godr. Fl. de Fr. i. 13; Woods, Tour. Fl. p. 3.

Habitat.—(A.) Hyères, collected by Dr. Shuttleworth;* flowers March 31st, fruit May 6th, 1868. (B.) Pegli, near Genoa; flowering specimens sent by Mrs. Tebbs, February 27th, 1867; fruit gathered by me at Savona April 17th, 1868. (C.) Mte. Mangiabo, near the Col de Broïs (Alpes Maritimes), collected by my father April 25th, 1867.

Remarks.—Anemone palmata, Linn., is one of the many rarities which Hyères and its neighbourhood possess. This species is not found anywhere else along the line of coast from Marseilles to Genoa (the

---

* I am glad to have so early an opportunity of thanking Dr. Shuttleworth for his great kindness in sending me boxes of plants by post for drawing. My best thanks are also due to Mrs. Tebbs, Dr. Bornet and M. Thuret, Dr. Hartsen, and others, for similar generosity.

tract of country to which the present work refers), but grows in Portugal (Welwitsch!), Andalusia (Nyman), Algiers (Schimper! Durien!), Sicily, and Sardinia (Nyman). It is interesting to observe that this plant, though most closely related to A. stellata, Lam., has very frequently a branched flower-scape, bearing a second flower and involucre. Some authors seem to consider this as a monstrous growth; but as the branched scape is the rule, and not the exception, in all strong and well-developed plants, this can scarcely be maintained. Several of the specimens in the British Museum Herbarium and in that at Kew exhibit this feature: in the latter collection, out of 24 specimens, 11 present a branched scape and 13 a simple scape, like that of A. stellata, Lam. Of course as the branch is always developed later than the primary flower, some of the 13 unbranched specimens might subsequently have become branched.* Indeed, we may suspect that whenever we see an unbranched scape in A. palmata, Linn., there is latent in that plant the power to produce a branch and second flower. This brings to mind the interesting subject of *latent characters*, so clearly discussed by Mr. Darwin in his last great work.† At page 61 Mr. Darwin, after having described a great number of cases in which long-hidden characters have reappeared in individuals, and thus revealed a forgotten or unsuspected descent, when referring to the wonderful nature of the germ which contains all these tendencies, says, " But on the doctrine of reversion as given in this chapter, the germ becomes a far more marvellous object; for, besides the visible changes to which it is subjected, we must believe that it is crowded with invisible characters, proper to both sexes, to both the right and left side of the body, and to a long line of male and female ancestors separated by hundreds or even thousands of generations from the present time; and these characters, like those written on paper with invisible ink, all lie ready to be evolved under certain known or unknown conditions."

A nemone trifolia, Linn., is a species most nearly allied to A. nemorosa, Linn., but is easily distinguished by its very regularly toothed leaves, while those of our English plant are irregularly cut and toothed like those of A. ranunculoides, Linn. This species replaces A. nemorosa, Linn., along the shore from San Remo to Genoa, this latter species usually growing higher up among the mountains, though I have found one small patch of plants of A. nemorosa, Linn., in the Varena Valley, near Pegli. I do not know of any habitat for A. trifolia, Linn., west of San Remo, but I suspect that it is to be found in the Tenda district.

A. ranunculoides, Linn., abounds amongst the mountains north of Mentone and on Mte. Ceppo and Mte. Bignone, near San Remo.

Explanation of Plate LI. *bis.*—Fig. A 1, immature achene of natural size. A 2, the same, magnified. Fig. B 1, head of fruit of natural size. B 2, one of the achenes of natural size. B 3, the same, magnified. Fig. C 1, head of fruit, magnified.

* Anemone palmata, γ. albida, figured and described by Sims in the 'Botanical Magazine,' t. 2079 (1819), is represented as having a branched scape.

† 'Animals and Plants under Domestication,' ii. pp. 50–61.

1.

2.

3.

PLATE LII.

DELPHINIUM Ajacis, Linn.

Natural Order RANUNCULACEÆ.

GEN. CHAR.—*Sepals* 5, subconnate at base, the uppermost one (or the calyx-tube) prolonged at the base outside into a spur. *Petals* 2-4, small, the 2 posterior (often united) prolonged into a spur within the calyx-spur, the 2 lateral not spurred or wanting. *Carpels* 1-5, sessile, free, containing many ovules, opening follicularly when ripe. *Leaves* alternate, subternately lobed or dissected. *Flowers* blue purple pink or white, very rarely yellow. Bth. et Hook. Gen. Plant. i. 9.

SPEC. CHAR.—*Flowers* large, in a lax, spike-like raceme; lower *bracts* similar to the leaves; *peduncles* only as long as or scarcely longer than sepals. *Follicle* pubescent, 2-3 times as long as broad, gradually tapering into the style. *Seeds* covered with continuous wavy ridges. *Leaves* ternately multifid. *Plant* annual.

Delphinium Ajacis, Linn. Sp. Pl. p. 748; Gren. et Godr. Fl. de Fr. i. 46; Woods, Tour. Fl. p. 9.

HABITAT.—Near Sospello (Alpes Maritimes). Collected by my father, Nov. 27, 1865.

REMARKS.—Delphinium Ajacis, Linn., closely resembles both D. Consolida, Linn., and D. pubescens, DC. D. Consolida, Linn., is much fewer and smaller flowered, and has long peduncles spreading stiffly and angularly, the bracts of one or two lobes only, a small glabrous capsule not tapering into the style, and seeds covered with interrupted ridges. D. pubescens, DC., has the bracts and habit of D. Consolida, Linn., but narrow, oblong-lanceolate sepals, and a pubescent capsule. It must be owned, however, that D. pubescens, DC., forms a link between D. Consolida, Linn., and D. Ajacis, Linn.; so that, should other intermediates be discovered, all three would have to be thrown into one species.

I have not been able to find any mention of habitats for D. Ajacis, Linn., between Nice and Marseilles. M. Ardoino* only cites it as growing at Castillon and Sospello, towns a few miles north of Mentone; and De Notaris,† when speaking of the coast-line from Nice to Genoa, gives no distinct habitats, but says that it is "common in the region of olives."

EXPLANATION OF PLATE LII.—Fig. 1, capsule, magnified. Fig. 2, seed, magnified. Fig. 3, the same, of natural size.

* Ardoino, 'Flore des Alpes Maritimes,' p. 20.

† De Notaris, 'Repertorium Floræ Ligusticæ in Memorie della Accademia della Scienze di Torino,' ser. 2, tom. viii. (1846), p. 9.

D.3.

D.1.

D. D.2.

D.4.

A.2.

A.3. A.

C.

C.3.

C.2.

C.1. C.4.

B.6.

A.1.

A.4.

B.

B.4.

A.5.

B.2.

B.5.

B.1.

B.3.

PLATE LII. *bis.*

(A) NIGELLA GARIDELLI, Moggridge; (B) N. ARVENSIS, Linn.;
(C) N. DAMASCENA, Linn.; (D) N. SATIVA, Linn.

Natural Order RANUNCULACEÆ.

GEN. CHAR.—*Flowers* terminal. *Sepals* 5–8, of regular shape. *Petals* irregular in shape, the limb being bifid and two-lipped. *Stamens* usually many (only about 10 in N. orientalis, Linn., and N. Garidelli, Moggridge), very regularly superposed in linear series, forming bundles which alternate with the petals (see Fig. B. 1), at first upright and then horizontally depressed; anther-valves horny, not shrivelling, the posterior pair larger, folded back at dehiscence so as to touch one another. *Carpels* 3–10, opening at the apex when ripe. *Plants* annual

(A.) SPEC. CHAR.—*Flowers* small. *Sepals* oblong-lanceolate, covered with minute white scales, upright. *Petals* opposite to and twice as long as sepals, lower lip of two linear lobes, without tubercles, covere below with white, clavate hairs, upper lip very short, ovate. *Carpels* 2–3, united almost in their whole length, covered with granular tubercles; *stigmas* almost sessile. *Leaves* trifid or pinnatifid, of few linear segments.

Nigella Garidelli, Moggridge. *Garidella Nigellastrum*, Linn. Sp. Plant. p. 608; Gren. et Godr. Fl. de Fr. i. 42; Woods, Tour. Fl. p. 8.

(B.) SPEC. CHAR.—*Flowers* not surrounded by leaves. *Sepals* 5–8, cordate-ovate, suddenly contracted below into a claw half its length, strongly 3-nerved. *Petals* not half as long as sepals, lower lip of two ovate lobes, attenuate above and ending in thickened knobs, hairy below, and having two small shining tubercles* on either side, near the cavity under the upper lip; upper lip ovate, attenuate into a filiform tip. *Anthers* apiculate. *Carpels* free in two-thirds of their length, glabrous; *styles* long, and spirally twisted, as in the two following species. *Leaves* pinnatifid, segments glabrous, linear-filiform.

Nigella arvensis, Linn. Sp. Plant. 753; Gren. et Godr. Fl. de Fr. i. 43; Woods, Tour. Fl. p. 8.

(C.) SPEC. CHAR.—*Flower* surrounded by leaves simulating an involucre. *Sepals* ovate, claw very short. *Petals* short, lobes of lower lip rounded, ciliate, furnished with two shining tubercles below; upper lip ovate, acute, rolled inwards into the deep cavity at the base of the limb. *Carpels* 4, or, more usually, 5, completely united, subglobular and swollen when ripe; the inner surface of each carpellary leaf separated so as to form a membranous inner cell round the ovules (Fig. C 4). *Leaves* tripinnatifid, the segments linear or linear-lanceolate in the lower leaves.

---

* By an oversight, these tubercles were omitted in the final drawing for Fig. B 3.

Nigella damascena, Linn. Sp. Plant. p. 753; Gren. et Godr. i. 43; Woods, Tour. Fl. p. 8. *Erobathos damascenum*, Spach, Suites à Buffon, vii. 301.

(D.) SPEC. CHAR.—*Flowers* free from leaves. *Sepals* ovate, claw very short. *Petals* short, lobes of lower lip ovate, attenuate, thickened at tip, furnished below with two shining tubercles and white, clavate hairs; upper lip ovate acute. *Anthers* exapiculate. *Carpels* completely united. *Leaves* irregularly tripinnate, segments lanceolate, hairy, hairs glandular-fusiform at base. *Stem* hairy.

Nigella sativa, Linn. Sp. Plant. p. 753; Gren. et Godr. i. 43; Woods, Tour. Fl. p. 8.

HABITATS.—(A, B, and D.) Cultivated specimens from Kew Gardens. (C.) Gathered by me at Mentone, April 30, 1867.

REMARKS.—Nigella Garidelli has usually been considered as belonging to the distinct genus Garidella; but, as Messrs. Bentham and Hooker point out,[*] the subsessile stigmas and small size of the flowers do not suffice to constitute a generic separation. I have therefore ventured, though most unwillingly, to introduce a new specific name into our already too-complex nomenclature, in order to avoid the unnatural severance which thrust this plant away from its own nearest blood-relations. Nigella Garidelli, Moggridge, grows near Marseilles and at Cassis, a short distance east of Marseilles (Roux!), at Toulon (Gren. et Godr.), formerly at Nice, and it has been stated that it grows at Cannes (Hanry). It is cited as coming from Granada, Crete, and Asia Minor (Nyman). In the quaint forms assumed by the petals in all the species of Nigella we are reminded of the well-known Venus-chariot in the flower of Monkshood (Aconitum), the irregular petals of Delphinium, and the horn-shaped petals of Hellebore and Winter Aconite (Helleborus and Eranthis). In fact, there is an evident tendency in all the members of this division of the great Ranunculus family, known as the Tribe Helleboreæ, to change the petals into small cups or pouches full of nectar.

Nigella arvensis, Linn., grows at Marseilles (Castagne and Derbes), and at Toulon (Gren. et Godr.). It has a very wide range through Central and Southern Europe, through Germany, France, Spain, Italy, Hungary, Croatia, Dalmatia, Bulgaria, to the Russian shores of the Black Sea (Nyman).

Nigella damascena, Linn., abounds as a weed of cultivation along the whole shore from Marseilles to Genoa. I have observed a form at Hyères which differs from that with which I am familiar at Mentone in the crowding and compactness of the more finely-divided leaves and the brighter colour of the sepals. This I have provisionally named var. conferta.

Nigella damascena, Linn., is exclusively South European, and ranges

[*] Gen. Plant. i. 8.

through Spain, Mediterranean France, Italy, Dalmatia, Greece, and Turkey, into Asia Minor (Nyman).

Nigella sativa, Linn., is said to grow at Grasse, near Cannes (Baudot in Gren. et Godr.), and has, generally speaking, a similar range to that of N. arvensis, Linn. It is cultivated for the sake of its aromatic seeds. Most of the species of the Nigella have the stamens placed in bundles of about seven each; these are at first upright around the ovary then curved outwards, and finally depressed horizontally when the anthers open their curious horny valves and the pollen is shed.

EXPLANATION OF PLATE LII. *bis.* Fig. A 1, flower, magnified. A 2, sepal, magnified. A 3, petal, magnified. A 4, fruit of natural size. A 5, carpels from a flower, magnified. Fig. B 1, a flower deprived of sepals to show the bundles of stamens placed singly in rows, as they appear when the pollen is shed. B 2, sepal of the natural size. B 3, petal, magnified (the two shining tubercles, one on either side of the hollow under the upper lip, have been accidentally omitted). B 4, an anther before dehiscence. B 5, an anther dehiscing. B 6, fruit of natural size. Fig. C 1, sepal of natural size. C 2, petal, magnified. C 3, fruit of natural size. C 4, lower half of the fruit, showing the double cells in each carpel. Fig. D 1, sepal of natural size. D 2, anther, magnified. D 3, fruit of natural size. D 4, the anatropous ovule, magnified.

1.

PLATE LIII.

CORYDALIS solida, Hook.

Natural Order PAPAVERACEÆ.

GEN. CHAR.—*Sepals* 2 (sometimes wanting), minute, like those of *Fumaria*. *Petals* 4, irregular, of the two outer ones the uppermost has a pouch or spur at the base. *Pod* many-seeded. *Seeds* lenticular.

SPEC. CHAR.—*Flowers* in a terminal raceme; *bracts* digitately lobed; *pedicels* (in specimens) shorter than the pods by one-half or one-third ("pédicelles aussi longs que la capsule," Gren. et Godr.). *Upper petal* emarginate. *Style* angularly bent during flowering. *Leaves* biternately cut, segments lanceolate. *Stem* furnished with a scale below the leaves. *Rootstock* tuberous, solid, subglobular.

Corydalis solida, Hook. Brit. Fl. (ed. 4) i. 265; Gren. et Godr. Fl. de Fr. i. 64; Woods, Tour. Fl. p. 13. *C. digitata*, Pers. Syn. Plant. ii. 269. *C. bulbosa*, DC. Fl. Fr. iv. 637.

HABITAT.—Collected by my father on Mount Aggel, near Mentone, April 6th, 1865.

REMARKS.—Corydalis solida, Hook., is a plant closely allied to C. fabacea, Pers. This latter, however, has a raceme of small flowers, which are almost hidden among the bracts, and, as its name imports, has broad leaf-lobes, each of which is almost as broad as long, and resembles the leaflet of the Bean.

I am indebted to the Rev. Wm. Hawker for a specimen of C. fabacea, Pers., from the Monte Rotondo in Corsica. This is a plant which should be searched for among the mountains of the Riviera. Our Mentonese form of C. solida, Hook., appears to differ from the type described by Grenier and Godron, Koch, and others, in its much longer pod.

The structure of the flower of Corydalis is rather complex, but is one adapted for insect fertilization, as the spur contains an attractive deposit of nectar.

Mr. Darwin* gives the following account of C. cava, Schweigg. (C. tuberosa, DC.):—

"Corydalis tuberosa properly has one of its two nectaries colourless, destitute of nectar, only half the size of the other, and therefore, to a certain extent, in a rudimentary state; the pistil is curved towards the perfect nectary, and the hood, formed of the inner petals, slips off the pistil and stamens in one direction alone, so that, when a bee sucks the perfect nectary, the stigma and stamens are exposed and rubbed against the insect's body. In several closely-allied genera, as in Dielytra, etc., there are two perfect nectaries, the pistil is straight, and the

* 'Animals and Plants under Domestication,' ii. 58.

hood slips off on either side, according as the bee sucks either nectary. . . . Now I have examined several flowers of C. tuberosa, in which both nectaries were equally developed and contained nectar, and in this we see only the redevelopment of a partially aborted organ; but with this redevelopment the pistil becomes straight and the hood slips off in either direction; so that these flowers have acquired the perfect structure, so well adapted for insect agency of Dielytra and its allies."

A similar case is mentioned by Hornung,[*] in which C. pumila, Host., presented two spurs, instead of one only, and thus resembled Dicentra and Adlumia, which always have two.

EXPLANATION OF PLATE LIII.—Fig. 1, the ovary taken from a flower, showing the angularly bent style.

* Bot. Zeit. (1836) xix. 2. p. 667.

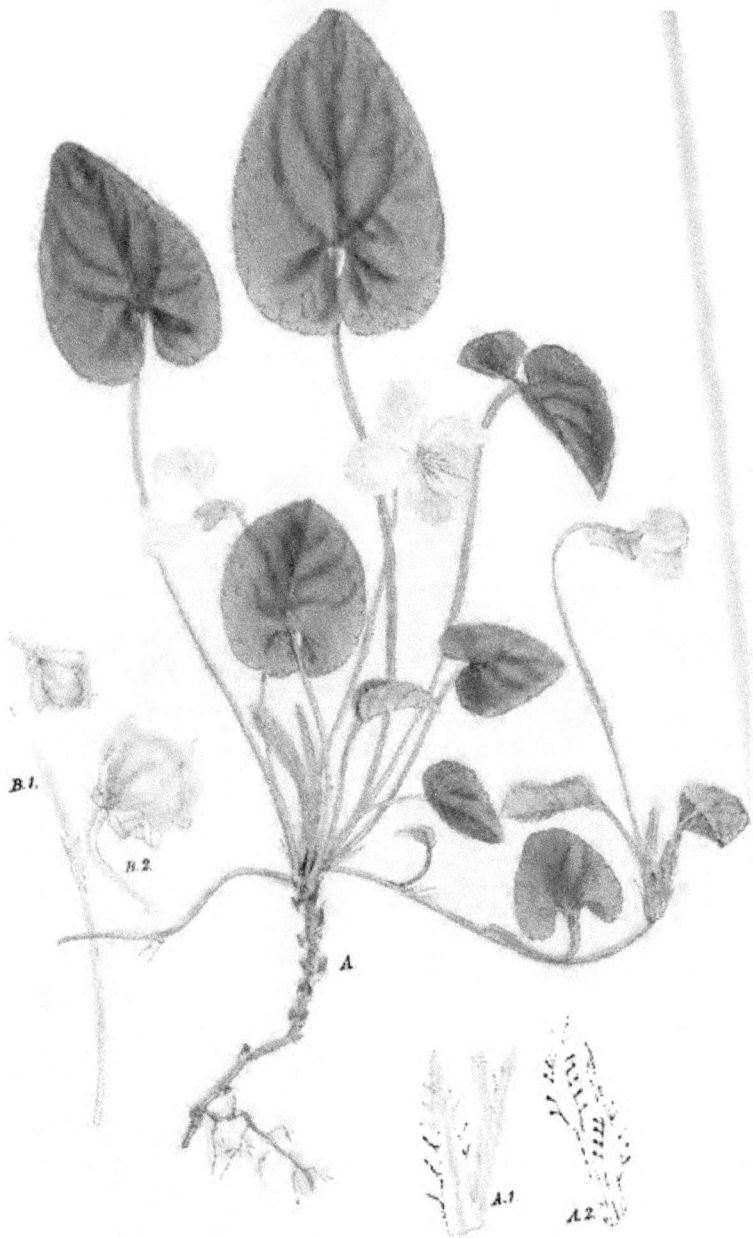

B.1.

B.2.

A.

A.1

A.2

PLATE LIV.

VIOLA HIRTA, Linn., var. *picta*.

Natural Order VIOLACEÆ.

GEN. CHAR.—*Sepals* nearly equal, prolonged at base beyond their insertion. *Petals* spreading, the lower one usually larger, having a spur or pouch at base. *Anthers* subsessile, connective flattened, prolonged into a membranous apex; two lower stamens usually spurred at the back. *Style capitate*, clavate or variously expanded, nearly straight with the stigma terminal, or more or less recurved with the stigma in front. *Capsule* opening in 3 valves. *Seeds* ovoid-globose; testa horny, often shining. Herbs, rarely undershrubs. *Leaves* alternate; *stipules* persistent, often leafy. *Peduncles* axillary, 1-, rarely 2-flowered. *Plants* (except in the Melanium section) producing flowers of two kinds, the earlier ones perfect, and often sterile; the summer flowers minute, having no petals or concealed petals, but more fertile. Benth. et Hook. Gen. Plant. i. 117.

SPEC. CHAR.—*Flowers* white, streaked with lilac, orbicular in outline, scentless; *peduncle* glabrous; *bracts* pubescent on back and edges in upper half; cilia glandular. *Sepals* pubescent at edge, obtuse. *Petals* broad, limb of the four upper ones broadly obovate or suborbicular, the two lateral being indistinctly emarginate at apex. *Capsule* hairy, globular, depressed. *Leaves* (when fully developed) ovate, prolonged, surface dull, coarsely pubescent, teeth long, straight along the back; *petiole* hairy, hairs deflexed; *stipules* linear, triangular-lanceolate, pubescent at edge and on back, cilia pubescent, sometimes glandular. *Stolons* short or wanting, not rooting.

Viola hirta, Linn. Sp. Plant. p. 1324, var. *picta*, Moggridge.

HABITAT.—(A) Bosco del Inferno, Albenga, specimens gathered by my father, March 16, 1868; (B) fruit gathered in the same spot by myself, April 23, 1868.

REMARKS.—This very pretty variety of Viola hirta, Linn., is established in such profusion in the Bosco del Inferno at Albenga as to constitute there a predominant race, having all the appearance of a distinct species; yet I believe it to be no more than a variety of the Hairy Violet, which is itself a rather doubtful species. In like manner there is at San Romolo, near San Remo, a colony of plants having pinkish flowers and distinctive characters in the sepals and stipules, which is also a variety of Viola hirta, Linn. The varieties of V. hirta, Linn., and of V. odorata, Linn., are most complex and confusing in the south, but careful observation will show that several prominent forms may be distinguished, and that these again may be divided into sub-varieties and local races. For example, at Mentone we have two predominant sub-species belonging to Viola odorata, Linn., the one having dark purple-blue flowers of an orbicular outline, hairy capsules, and very narrow, linear-triangular stipules (V. scotophylla, Jord.); the other, pale lilac-blue flowers, with large white eye, of an oblong outline, glabrous fruit, and broad, lanceolate stipules (V. floribunda, Jord.). But, from close acquaintance with the Mentone district, I could show several colonies of Viola floribunda, Jord., which are distinguishable from one another, and also of V. scotophylla, Jord.; these being varieties and sub-varieties belonging to either sub-species. Most species of Viola, with the exception of those belonging to the Melanium or Pansy section, have two kinds of flowers, which succeed one another on the same plant; the spring flowers are those with which we are all so familiar, but the summer flowers are minute, scarcely bigger than a pin's head, and easily to be mistaken for small flower-buds,—and yet these are the ones which produce the greatest profusion of seed! One may observe flowers in the early spring which have curiously distorted, coloured flowers, and in the late spring others which, though coloured and perfect, are mere miniature Violets of a very small size; these are intermediate stages between either condition.

EXPLANATION OF PLATE LIV., Fig. A 1, stipules, magnified. A 2, bracts and part of peduncle, magnified. Fig. B 1, fruit of natural size. B 2, the same, magnified.

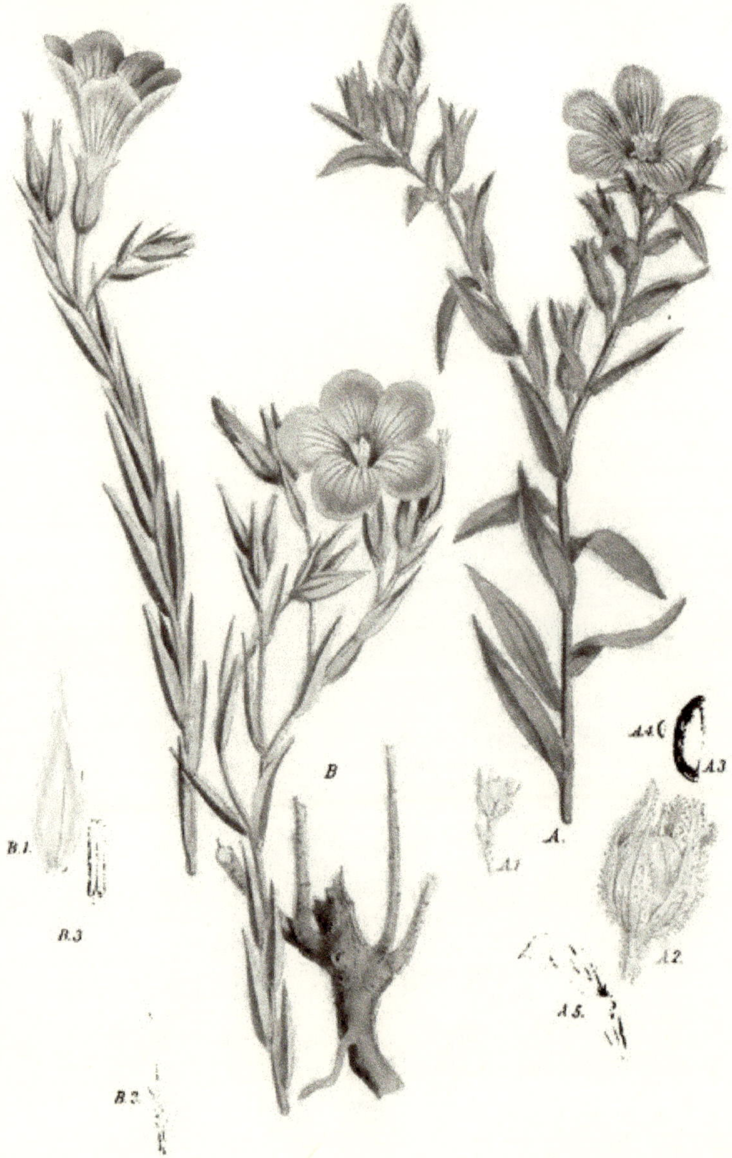

Vincent Brooks Day & Son.

PLATE LV.

(A) LINUM viscosum, Linn.; (B) L. narbonense, Linn.

_____

Natural Order LINACEÆ.

GEN. CHAR.—See description of Plate V. Part I.

(A.) SPEC. CHAR.—*Flowers* large, pink, veined with purple, dimorphic, in scorpioid or corymbose racemes; *peduncles* half as long as sepals. *Sepals* ovate, alternate, densely glandular, pubescent, longer than capsule. *Petals* beautifully veined, cuneate-obovate, claw yellow, dilated into a prominent ridge on inner side. *Capsule* somewhat 3-ribbed, distinctly apiculate. *Leaves* ovate-lanceolate, 3-5-nerved, glandular, pubescent or ciliate, without glands at base. *Stem* glandular, pubescent, woody below, perennial.

Linum viscosum, Linn. Sp. Plant. p. 398; Gren. et Godr. Fl. de Fr. i. 281; Woods, Tour. Fl. p. 61.

(B.) SPEC. CHAR.—*Flowers* dimorphic in a corymb; the young corymb of buds and upper part of the stem nodding, fruiting; *peduncles* erect. *Sepals* glabrous, membranous at edge, lanceolate-acuminate. *Petals* obovate, on long claw. *Anthers* oblong, about three times as long as broad. *Capsule* shorter than calyx. *Leaves* lanceolate, glabrous, minutely denticulate at edge, without glands at base. *Stems* many, from a perennial woody stock.

Linum narbonense, Linn. Sp. Plant. p. 398; Gren. et Godr. Fl. de Fr. i. 282; Woods, Tour. Fl. p. 60.

HABITATS.—(A.) Gorbio Valley, Mentone, gathered December 18, 1866; fruit drawn from a specimen in M. Ardoino's herbarium, without habitat given, but probably from the same spot. (B.) Gathered by me at Oneglia, on April 11, 1867.

REMARKS.—*Linum viscosum*, Linn., is one of the many plants which linger on in flower until the winter, and may still be found in November and December. This species is not rare along the coast from Genoa to Nice, but is principally found in the submountainous region. Bertoloni mentions having gathered it near Genoa ("Allo Sperone"). It is a very ornamental plant, and well merits introduction into our gardens. I do not know of any habitat for this species to the west of Nice until one reaches the Vallée de Castanese in the Pyrenees (Fourçade!) Nyman, in his 'Sylloges Floræ Europæ,' records it as growing among the Eastern Pyrenees, in Arragon, Navarre, and Portugal, in Italy, Austria (Tyrol), and Germany (Oberbayern). Linum narbonense, Linn., abounds on the hills near Oneglia, among the mountains north of Nice and Mentone, often descending along the sides of the valleys, on Mount Coudon and its slopes, near Hyères, and at Marseilles (Castaigne et Derbes).

EXPLANATION OF PLATE LV.—Fig. A 1, a fruit of the natural size. A 2, the same, magnified. A 3, a seed, magnified. A 4, the same, of natural size. A 5, claw of a petal, magnified. Fig. B 1, sepal, magnified. B 2, petal, of natural size. B 3, stamen, magnified.

PLATE LVI.

LINUM CAMPANULATUM, Linn.

Natural Order LINACEÆ.

GEN. CHAR.—See description of Plate V. Part I.

SPEC. CHAR.—*Flowers* yellow, dimorphic, in a loose corymb; *peduncles* very short. *Sepals* lanceolate-acuminate, the membranous edge ciliate, some of the cilia glandular. *Petals* obovate, on a long claw, mucronulate at apex. *Capsule*, valves acuminate, shorter than calyx. *Leaves* glabrous, lanceolate above, then cuneate-lanceolate, and finally spathulate below, having the transparent membranous edge finely denticulate or ciliate; there is a gland on either side at the base of the leaf.

Linum campanulatum, Linn. Sp. Plant. p. 400; Gren. et Godr. Fl. de Fr. i. 280; Woods, Tour. Fl. p. 61.

HABITAT. — River bed near Drap, north-east of Nice, where I gathered it on May 11, 1867.

REMARKS.—Judging from dried specimens, it is rather difficult to separate as a distinct species Linum campanulatum, Linn., from L. flavum, Linn. Linum flavum, Linn., has a more compact inflorescence, and, according to Reichenbach (Ic. Fl. Germ. vi. tab. 340, 341, p. 835), is amply distinguished also by the petals, which are rounded, and not mucronulate at apex, by the shorter sepals and less acuminate fruit.* There is, moreover, a peculiar aspect which would perhaps enable one always to distinguish typical specimens of either plant. I should not, however, be surprised to find that intermediate forms exist, for the two species are certainly most closely related. The geographical range of L. flavum, Linn., is more easterly than that of L. campanulatum, Linn., the latter being found in Spain, Southern France, Italy, and Dalmatia (Nyman), while L. flavum, Linn., is found in Austria, Germany (Ulm), Lombardy, Croatia, Hungary, Transylvania, Turkey (Thrace), and on the Russian shores of the Black Sea. I have gathered L. campanulatum, Linn., without flowers, at Pegli, near Genoa. I have complete specimens also from Varazze, Arenzana, and from Hyères (Mont de Paradis, gathered by Dr. Shuttleworth).

EXPLANATION OF PLATE LVI.—Fig. 1, leaf and stem, cut about midway, showing the glands at the base of the leaf and the prominent ridges of the stem. Fig. 2, a sepal. Fig. 3, a petal of the natural size. Fig. 4, a capsule. All the Figures magnified, except Fig. 3.

* Bertoloni points out that even the lower leaves are never spathulate, as they are in Linum campanulatum, but are all pointed and obversely-lanceolate.

Vincent Brooks Day & Son, Imp

PLATE LVII.

ADENOCARPUS TELONENSIS, Gay.

Natural Order LEGUMINOSÆ.

GEN. CHAR.—" *Calyx* persistent, often glandular, bilabiate, upper lip bipartite, lower lip longer, trifid, the central lobe longer than the lateral lobes. *Petals* connate at the base with the filament-tube ; *keel* obtuse, covering the stamens and style, limb of keel usually furnished on either side with a conical gibbus. *Stamens* monadelphous, filaments alternately longer, anthers of the shorter stamens linear, of the longer ovoid. *Pod* linear, plano-compressed, somewhat thickened, stiff, dilated at apex, glandulose-tuberculate all over (*undique echinatum*). *Shrubs* with divaricate branches and white bark. *Leaves* trifoliolate, usually having axillary, rudimentary, leafy branches in the axils. *Flowers* yellow, usually racemose, very seldom capitate. *Pedicels* furnished with two bracts at the middle or below. *Seeds* blackish, ovate, somewhat compressed, furnished with a perisperm ; perisperm rather thick, horny, separable from the thin integument (*in seminibus maceratis*)." Gay, Adenoc. Monog. MSS. ined.[*]

SPEC. CHAR.—*Flowers* yellow,[†] capitate, 1-3, terminating the branches; *bracts* often in pairs or threes, ovate-acuminate, ciliate. *Calyx* hairy, not glandular, lower lip as long as keel. *Standard* covered on the back with silky hairs, yellow (yellow mixed with purple-brown, Gay†); *keel* without conical bosses on either side. *Pod* compressed, wavy, oblong, covered with short, clavate, or cylindrical warts. *Leaves* of 3 small obovate leaflets, glabrous, or having on either face a few long white hairs. *Growth* shrubby, about 3 feet high, branches complicated, somewhat resembling that of the common thorny Cytisus (*C. spinosus*, Lam.).

Adenocarpus telonensis, Gay, Adenocarpi Monog. ined. *A. grandiflorus*, Boiss. Bibl. Un. Gen. (1866), et Voy. en Esp. i. tab. 42, descr. ii. 146.

HABITAT.—Hyères, in pine-woods near the town, where I gathered it, May 13, 1867.

REMARKS.—The genus Adenocarpus is easily distinguished from its allies by its tuberculate fruit. The synonymy of the species has, how-

---

[*] I am indebted to Dr. Hooker for permission to transcribe from the valuable manuscripts of M. J. Gay, which are now in his possession. The above is a translation from the Latin. No date is affixed to the MSS.

[†] Gay, in describing A. telonensis, Gay, from specimens gathered at Hyères, gives the colour of the flowers as yellow, mixed with purple-brown. Boissier also, in his coloured figure of this species from Malaga specimens, represents the flowers of a dingy brown and yellow. However, I cannot think that when drawing the flowers from fresh specimens I could have mistaken the colour.

ever, been brought into great confusion, and it was not until I had discovered a monograph of this genus among Gay's unpublished writings that I clearly saw my way. I think it well, therefore, to publish an abstract of the descriptive portion of this monograph, with the synonymy and geographical distribution.

"1. Adenocarpus anagyrius, N. A. elatus, grandifolius, stipulis villosissimis, floribus racemosis, bracteis . . ., carina . . . legumine multiglanduloso.—C. anagyrius, L'Her. Stirp. p. 184 (excl. syn. Clus. et C. Bauh.). Adenocarpus hispanicus, DC. Fl. Fr. Suppl. p. 549, Prodr. ii. 158. A. anagyrius, Spreng. Syst. Veg. iii. 226 (excl. patria). —Habitat in umbrosis Hispaniæ, speciatim montibus del Paular (Herb. Desf.) et in Lusitaniâ. Differt . . . ab omnibus sequentibus statura multo altiore, caule valido et fere arborescente, dupla foliolorum et seminis magnitudine, legumine dimidio latiore, et perispermine in seminibus nullo.

"2. Adenocarpus complicatus, N. A. humilis, parvifolius, stipulis glabriusculis, floribus racemosis, bracteis 2 lineari-lanceolatis, calyce erecto, carina bigibbosa, legumine multiglanduloso.—Spartium complicatum, Linn. Sp. ed. i. p. 709.

"α. polyadenius. A. ramis rachique glabriusculis, floribus majusculis, calyce glabriusculo multiglanduloso, glandulis nigricantibus, seminibus ovato-oblongis, immaculatis.—Cytisus divaricatus, L'Her. Stirp. p. 184 (excl. syn. Sauv.); Ait! Hort. Kew. ed. 1, iii. p. 50. C. parvifolius, Lam. Dict. ii. p. 248. Spartium complicatum, Thor. Chlor. Land. p. 297. C. complicatus, DC. Fl. Fr. n. 3821. Adenocarpus parvifolius, DC.! Fl. Fr. Suppl. p. 550; Prodr. ii. p. 158. Cytisus nigricans, Loret et Duret! Cat. Pl. Côte d'Or (1825), p. 27.—Habitat in Gallia occidentali, ab Aturo ad Ligerem, pluribus locis; in agro Santonico (Lois); in Pictavia, . . . Citatur præterea in Etruria, . . . in collibus circa Romam, . . . et in Eubœa . . .

"β. hirsutus. A. ramis villoso-hispidis, rachi villoso-pubescente, floribus majusculis, calyce pauci-glanduloso eglanduloso-ve, villoso, seminibus ovatis immaculatis.— Cytisus divaricatus, β. Sims, Bot. Mag. tab. 1387. C. divaricatus, Ten. Fl. Nap. ii. 147. Adenocarpus commutatus, Guss. Fl. Sic. Prodr. ii. (1828), p. 375.—Habitat in Calabriæ monte Mongiana (L. Thom.), in arenosis montosis Siciliæ, Messina ulla portella Ibiso Monte Scuderi (Guss.); in Lusitania, . . . in Hispania centrali circa Salmanticam.

"γ. anadenius. A. ramis rachique glabriusculis, floribus mediocribus, calyce eglanduloso villosiusculo, seminibus ovatis maculatis.— Spartium complicatum, Gouan, Hort. Monsp. p. 356. Adenocarpus telonensis, DC. Fl. Fr. Suppl. p. 550 (excl. syn. Ger. et Lois et loc. nat. Galloprov.); Prodr. ii. p. 158 (excl. iisd.).—Habitat in Cebennorum apricis declivibus, frequens, vix extra castancarum regionem (Alais, La Salle, Bois de Valène, La Sérane, inter Gauges et Sumene, etc.), . . . denique in præfectura De la Lozère.

"δ. oligadenius. A. ramis glabriusculis, rachi, pubescente calyce villoso-pubescente, pauciglanduloso eglandulosove, seminibus . . .—Cytisus complicatus, Brot. Fl. Lusit. ii. p. 158. Adenocarpus intermedius?, DC. Fl. Fr. Suppl. p. 549; Prodr. ii. p. 158.— Habitat in Lusitaniæ arenosis apricis, tum ad Mundam (Brot.), tum ad Durium, præsertim circa urbem Porto (Herb. Tournef., Vaill., Isn., Brot.).

"3. Adenocarpus foliolosus, DC. A. humilis, parvifolius, stipulis villosiusculis, floribus racemosis, bracteis 2 lineari-lanceolatis, labiis calycinis reflexis, carina bigibbosa, legumine parci-glanduloso.

"α. exadenius. A. calyce eglanduloso vel pauci-glanduloso.—C. foliolosus, L'Her. Stirp. p. 184. Ait! Hort. Kew. ed. 1, iii. p. 49. Adenocarpus foliolus, DC. Fl. Fr. Suppl. p. 549; Prodr. ii. p. 158. Genista foliolosa, Lk. Enum. Alt. ii. p. 224. Spartium microphyllum, Cav. ex Leop. von Buch, Phylic. Beschreib. der Canar. Inseln (1825), pp. 155 et 184.—Habitat in insularum Canariensium fruticetis et sylvis, speciatim in Canaria magna, et in Teneriffa circa urbem Laguna (DC).

"β. polyadenius. A. calyce multiglanduloso.—Adenocarpus frankenioides, Chois. in DC. Prodr. ii. p. 158.—Habitat in declivitate montis Teneriffensis, alt. 500 ped. et ab incolis Codeso dicitur (Christ. Smith, ex DC. Prodr.).

"4. Adenocarpus telonensis, N. A. humilis, parvifolius, stipulis glabriusculis, floribus paucis capitatis, bracteis 3 verticillatis ovatis acuminatis, calyce erecto, carina æquali legumine multiglanduloso.—Cytisus telonensis, Lois!, Fl. Gall. ii. p. 416.— Habitat in Galloprovincia circa Olbiam (Hyères) (Robert, Lois); in Corbariis (Pourr. e frustulo in herb. Lemon); in Hispaniæ collibus circa Malagam."

EXPLANATION OF PLATE LVII.—Fig. 1, a flower in bud, with pedicel and bracts, magnified. Fig. 2, the standard, magnified. Fig. 3, a pod of the natural size.

PLATE LVIII.

CYTISUS ARDOINI, Fournier.

Natural Order LEGUMINOSÆ.

GEN. CHAR.—*Calyx-teeth* or lobes short, two upper ones united into a lip or free. *Standard* suborbiculate or ovate; wings obovate or oblong; *keel* straight or incurved, obtuse or scarcely acuminate, claws free. *Stamens* all united into a closed tube; alternate *anthers* shorter, versatile, the longer ones attached by the base. *Ovary* sessile or rarely stalked, containing many ovules; *style* incurved, glabrous, the terminal *stigma* oblique or capitate. *Pod* compressed, flat, oblong or linear, glabrous or hairy, of two valves. *Seeds* strophiolate.* *Leaves* sometimes digitately 3-foliolate, sometimes having but one or no leaflet. *Stipules* minute. Benth. et Hook. Gen. Plant. ii. 484.

SPEC. CHAR.—*Flowers* yellow, 1–6 in the axil of each leaf, usually secund; *pedicels* about twice calyx, without bracteolæ, hairy. *Calyx* campanulate, scarious in upper half, hairy, lips divergent, upper lip entire or indistinctly bidentate, lower lip indistinctly 3-toothed. *Standard* orbicular, abruptly narrowed into a short claw, incurved at edges (and thus simulating many species of Genista), quite glabrous; *wings* about as long as standard; *keel* slightly shorter than wings, somewhat pointed, but not rostrate, glabrous, the two petals which form the keel free, except close to the claws, each having on the outer side, near the base of the limb, a conical prominence corresponding to a depression in the adjacent wing. *Pod* hairy, compressed, nearly flat on either side, oblong, attenuate at base into a stalk. *Leaves* 3-foliolate, leaflets obovate, hairy, small, silky when young. *Stems* rod-like, generally decumbent, many springing from a knotty and twisted stock.

Cytisus Ardoini, Fournier, Bull. Soc. Bot. Fr. xiii. (Comptes Rend. 1866), p. 389; Ardoino, Fl. Alp. Mar. p. 93.

HABITAT.—Mont Cima d'Ours, near Mentone, collected in flower only by my father, April 22, 1867; the pod (Fig. 6) gathered by the Rev. William Hawker in May, 1866.

REMARKS.—This very pretty, newly-discovered species is only known to grow on the summits of three mountains near Mentone, namely, Cima d'Ours, the Aiguille, and Mont de Meras. It has been named after the author of the 'Flore des Alpes Maritimes,' whose name is so intimately associated with the botany of the neighbourhood of Mentone.

---

* Messrs. Bentham and Hooker depend principally upon this character for the distinction between Cytisus and Genista. A strophiole or caruncle is an enlargement of the outer coat of the seed, forming a scar, wart, or other excrescence.

Cytisus Ardoini, Fournier, is closely related to Cytisus glabrescens, Sartorelli* (= C. emerifolius, Rehb.), but may be distinguished† by having hairy pods, while those of C. glabrescens, Sartorelli, are glabrous; the keel also of this latter plant appears decidedly rostrate. Cytisus glabrescens, Sart., differs, moreover, according to Bertoloni,‡ in having the leaves glabrous on the upper side, and the standard yellow, veined with red. Cytisus Ardoini, Fourn., suffers greatly from the ravages committed by the cattle, which are driven in herds from place to place, and eat leaves, twigs, and everything that is green and not poisonous. It is, therefore, extremely difficult to obtain fruit, and the only pods that I have seen are those placed by the Rev. W. Hawker in M. Ardoino's herbarium. When my father was among the mountains of the Alpes Maritimes in July last (1868), he made an excursion in search of pods to Mont de Meras, where he had, during the spring previous, seen the ground covered with its golden flowers; but the cattle had been before him, and so closely had they eaten everything down, that he failed to discover any characteristic portion of the plant which might lend to recognition. This probably happens every spring, and will account in part for the extreme rarity of this species, and perhaps also for its low and stunted growth.

EXPLANATION OF PLATE LVIII.—Fig. 1, the standard, showing the inner face and the incurved margins. Fig. 2, one of the wings, showing the depression corresponding to the conical boss on the keel. Fig. 3, the keel, showing one of the two conical bosses. Fig. 4, the flower deprived of its petals. Fig. 5, two of the stamens, one short, with ovate, versatile anther, the other long, with oblong anther attached by the base. Fig. 6, a ripe pod of the natural size. Fig. 7, the ovary, taken from a flower. All the Figures are magnified except Fig. 6.

* Degli Alberi indig. al Boschi del Ital. super. p. 282 (1816).

† I judge from specimens in the herbarium of the late M. J. Gay (recently purchased by Dr. Hooker, and presented by him to the Royal Herbarium at Kew), which are the only ones I have seen.

‡ Bertoloni, 'Flora Italica,' vii. 562. This species is only found near the Lake of Como and amongst the neighbouring Swiss mountains.

C

A

A.1

A.2

A.3

B

B.4

G

B.5

B.1

B.2

B.3

PLATE LIX.

(A) ERICA MULTIFLORA, Linn.; (B) E. ARBOREA, Linn.; (C) E. SCOPARIA, Linn.

Natural Order ERICACEÆ.

GEN. CHAR.—*Flowers* either axillary or in short, terminal clusters, mostly drooping. *Sepals* 4. *Corolla* ovoid, globular or campanulate (in some exotic species tubular), more or less 4-lobed, and persisting round the capsule till its maturity. *Stamens* 8. Capsule free, with 4 cells, opening in as many or twice as many valves, each cell with several seeds.—Much branched shrubs, usually low, but in some species attaining 8–10 feet, with small, entire leaves, usually in whorls of 3–4, but sometimes opposite or scattered, and almost always rolled back on their edges. Benth. Handbk. Brit. Fl. (1865), p. 526.

(A.) SPEC. CHAR.—*Flowers* bright pink, forming a terminal, obconic, centrifugal raceme; *peduncles* longer than the leaves; *bracteolæ* oblong-ovate, ciliate. *Calyx-lobes* lanceolate, nearly half as long as corolla. *Corolla* ovoid, prolonged, about one-third longer than broad. *Anthers* exserted, without appendages, divided nearly to their base into two linear-oblong cells, dehiscing by two pores occupying not more than one-third of their length; *filament* inserted rather above and at the back of the gibbous base of the anther. *Leaves* thick, and almost cylindrical, in whorls of 4 or 5. *Stems* erect, woody, one foot to 18 inches high.

Erica multiflora, Linn. Sp. Plant. p. 503; Gren. et Godr. Fl. de Fr. ii. 429; Woods, Tour. Fl. p. 242. *Erica multiflora longipedicellata*, Wendl. Eric. fasc. 5 (1799), p. 7 (bene quoad iconem, excl. patr. Lusit.). *E. vagans*, Desf.! Atl. (1800), i. 329. *E. peduncularis*, Presl, Delic. Prag. (1822), p. 89.*

(B.) SPEC. CHAR.—*Flowers* nearly white, in a long, branched, tapering, spike-like panicle, 6–18 inches long. *Calyx-segments* ovate, about one-third of corolla. *Corolla* globular, ovate, nearly as broad as long. *Stamens* included; anthers united nearly to apex, furnished at the back with two orbicular, denticulate appendages. *Stigma* broad, peltate, convex (having 4 tubercles in the centre, Gay, MSS.). *Leaves* in closely-packed whorls of 3 each. *Branches* hairy, hairs themselves hairy and

---

* Synonymy taken from Gay's unpublished MSS. on Erica (1832, in part), which include descriptions and copious notes of the characters of the following species :— Erica lugubris, Salisb.; (E. mediterranea, Auct.); E. carnea, Linn.; E. multiflora, Linn.; E. verticillata, Linn.; E. vagans, Linn.; E. ciliaris, Andr.

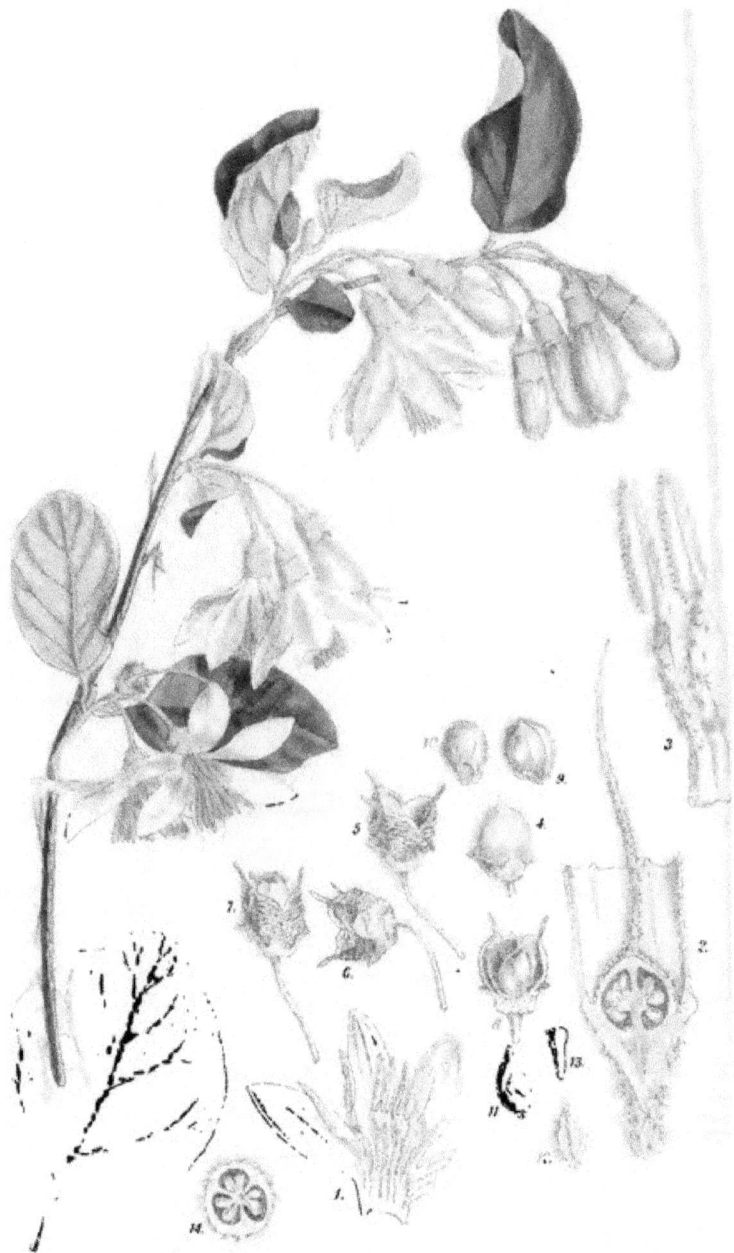

PLATE LX.

STYRAX OFFICINALE, LINN.

Natural Order STYRACEÆ.

GEN. CHAR.—*Calyx* urceolate, campanulate, 5-toothed or nearly entire. *Corolla* gamopetalous, 5-partite, rarely and perhaps abnormally (*monstrositate*) 4- or 6-7-partite. *Stamens* 10, rarely 7-12, united with the base of the corolla, opposite to and alternate with its lobes. *Filaments* united at the base into a short tube, distinct at the apex, hairy on inner side. *Anthers* erect, adnate, bilocular. *Ovary* adherent at base, ovoid, pubescent, 3-locular, the partial walls (*parietibus incompletis*) separating at length from the axis. *Ovules* many. *Seeds* usually solitary. *Albumen* fleshy. DC. Prodr. viii. 259.

SPEC. CHAR.—*Flowers* in small axillary or terminal cymes of 2-5 flowers each, nodding. *Calyx* tubular, dilated below at the point of adhesion with the ovary, having 5 minute teeth, 3-lobed in fruit. *Corolla* deeply 5-7-lobed, lobes lanceolate. *Stamens* 10-12, united at base, furnished with stellate hairs on the margins of the anther-cells and the filaments. *Fruit* composed of a pubescent, somewhat fleshy pericarp, which splits into three mucronate, wrinkled lobes, and leaves the shining orange-brown nut free within. *Leaves* ovate-orbicular, nearly glabrous on the upper side, densely covered with whitish, tomentose, stellate hairs below, as are the petioles, young branches, and peduncles. *Growth* that of a large shrub, 7-12 feet high.

Styrax officinale, Linn. Sp. Plant. p. 635 ; Gren. et Godr. Fl. de Fr. ii. 470 ; Woods, Tour. Fl. p. 244.

HABITAT. — Base of Mount Coudon, near La Farlede, north of Hyères, where I collected the flowering specimens figured, on May 7, 1868 ; the fruits represented (Figs. 5-12) were drawn from specimens kindly given me by Dr. Shuttleworth, collected at Montrieux ; and Fig. 4 from a specimen in the Kew Herbarium from Mount Olympus in Asia Minor (no. 2544 Aucher-Eloy, Herbier d'Orient).

REMARKS.—It is well worth while to make a journey to Hyères in the month of May, on purpose to see the bushes of Styrax in full flower near La Farlede. When arrived at La Farlede, which is about an hour's drive distant from Hyères, inquire for the cart-track which the inhabitants of the village know as the Pas-de-Galle, and follow this till, after about a quarter of an hour's walk, you reach a stream descending from Mount Coudon, where the banks are clothed with this beautiful shrub. Styrax officinale, Linn., is widely spread over Mount Coudon and the adjacent district towards Toulon and to the eastward. It has been said to grow in the forests near the river Var, but I have never met with any botanist who felt sure that it still exists there. The more adventurous visitors staying at Cannes should try to clear up this doubtful question. From Marseilles to Genoa, the only certain habitats for this plant are those in the more mountainous districts between Toulon and the Esterelles. Styrax officinale, Linn., yields a resinous balsamic juice, obtained by crushing the bark, which is known as storax, and used in perfumery and medicine. "Styrax Benzoin, a native of Sumatra, Borneo, etc., yields the resin called Benzoin. Benzoin is employed medicinally in chronic pulmonary disorders, and also by perfumers. It is used in Roman Catholic churches in the composition of incense."[*]

Explanation of Plate LX.—Fig. 1, a portion of the corolla and stamens. Fig. 2, calyx and ovary longitudinally divided. Fig. 3, two stamens. Fig. 4, a fruit before dehiscence of the pericarp. Figs. 5, 6, the fruit when the pericarp has split into 3 lobes and the nut has fallen out. Fig. 7, the same, with the nut inside. Fig. 8, the same, cut longitudinally. Figs. 9, 10, the nut showing its oblique base. Fig. 11, the seed. Fig. 12, the seed, longitudinally divided so as to show the embryo lying in the glutinous albumen. Fig. 13, the embryo. Fig. 14, a transverse section of the ovary. Figs. 1, 2, 3, 11, 12, 13, 14, magnified; Figs. 4, 5, 6, 7, 8, 9, 10, natural size.

[*] Lindley and Moore, 'Treasury of Botany,' ii. 1109.

PLATE LXI.

CONVOLVULUS sabatius, Viv.

Natural Order CONVOLVULACEÆ.

GEN. CHAR.—See description of Plate XXX. Part II.

SPEC. CHAR.—*Flowers* of a fine mauve colour, with small yellow centre, solitary or in pairs, crowded together towards the ends of the branches, two or three expanded at the same time on the same branch ; *peduncles* and *pedicels* covered with adpressed hairs (recurved when fruiting, Rchb.[*]) ; *bracts* linear or linear-lanceolate, distant from the flower. *Sepals* oblong-lanceolate, contracted under the apex, covered with small, white, adpressed hairs. *Anthers* white. *Disk* bright yellow, forming a prominent, cup-shaped ring round the ovary. (*Capsule* globose, nearly equalling calyx, 2-celled and 2-valved, Bert.[†]). *Leaves* thick in texture, opaque, set closely together, oblong-ovate, truncate or subcordate at the base, obtuse or emarginate, lowest leaves obovate, covered with minute, adpressed hairs ; *petioles* about one-third of limb. *Branches* decumbent, not rooting or twining, densely covered when young with minute, adpressed hairs, forming a dense mat round the central knotted perennial stem.

Convolvulus sabatius, Viv. Fl. Libyc. Spec. p. 67 ; Bert. Fl. Ital. ii. 442 ; Dutrs. Rep. Fl. Liq. i. 316 ; Woods, Tour. Fl. p. 250. *C. pseudo-siculus*, Cav., β. *multiflorus*, Choisy in DC. Prodr. ix. 407.

HABITAT.—Capo di Noli, eastward of Finale, where I gathered it on April 20, 1868.

REMARKS.—This very beautiful Convolvulus is only known to grow on this one promontory of Noli, the Vada Sabatia of the Romans, in the whole world. It is, therefore, necessary to consider its affinities with other species carefully. Undoubtedly its nearest ally is Convolvulus mauritanicus, Boiss.,[‡] a plant found near Constantine, in Algiers, and now largely grown in English gardens. Perhaps Convolvulus mauritanicus, Boiss., should be treated as a variety of Convolvulus sabatius, Viv., but I find from the description and drawing given by Boissier that it differs in the following characters:—Convolvulus mauritanicus, Boiss., has the flowers and leaves much scattered, and leaves, calyces, and younger parts of the stem covered with long, spreading hairs, the leaves are of a thin texture, the anthers also are represented as being yellow, and the corolla striped with pink along the folds. But we must remember that Convolvulus sabatius, Viv., only grows on this one great

[*] Ic. Fl. Germ. xvii. tab. 1337, fig. 2.   [†] Fl. Ital. ii. 442.
[‡] Voy. en Esp. ii. 418, et tab. 122 a, vol. i.

precipitous limestone cliff, almost overhanging the sea, west of the little village of Noli, so that its isolated position—and especially the fact that it is close to the sea—may account for some of its peculiarities. I cannot see any grounds for treating Convolvulus sabatius, Viv., as a mere variety of Convolvulus pseudo-siculus, Cav., as Choisy did in De Candolle's 'Prodromus,' as this latter plant has flowers scarcely one-third of the size of those of Convolvulus sabatius, Viv., and of white colour, tinged with pink and yellow at the centre, and the corolla superficially 5-lobed, besides other important differences. The cliff of Noli harbours several charming plants, foremost among which I may mention Campanula floribunda, Viv. Those who drive along the road between Finale and Noli in the autumn or early winter should not forget to look out for the beautiful purple-blue flowers of this plant. Now, both Campanula floribunda, Viv., and Convolvulus sabatius, Viv., are plants peculiar to Noli and its neighbourhood, whose nearest allies are now found far away to the south. Have their relations once been growing along the Riviera, and have they now fled southward? or are these the outposts of a new advance,—small colonies sprung from some far-travelled seed come from the old African or Sicilian home?

EXPLANATION OF PLATE LXI.—Fig. 1, a flower deprived of the stamens and corolla, magnified. Fig. 2, the same deprived of the calyx, showing the cup-shaped disk surrounding the ovary, magnified.

PLATE LXII.

(A, B) ORIGANUM vulgare, Linn., β. prismaticum, Gaud. ;
(C, D) CALAMINTHA Nepeta, Link et Hoffm.

Natural Order Labiatæ.

(A, B.) Gen. Char.—*Flowers* gathered into cylindrical or oblong spikelets. *Bracts* imbricate, coloured, covering the calyces. *Calyx* tubular (*ovato-tubulosus*), 10–13-nerved, striate, the 5 teeth equal or the upper ones scarcely longer; throat hairy within. *Corolla-tube* equalling or scarcely exceeding the calyx, limb subbilabiate, upper lip suberect, emarginate, lower lip spreading, trifid, lobes nearly equal. *Stamens* exsert, separated, the lower pair longer. *Achenes* dry, nearly smooth. *Leaves* quite entire or slightly lobed. Benth. Lab. p. 334.

(A, B.) Spec. Char.—*Flowers* in elongated spikelets, forming a loose, terminal corymb; the large hermaphrodite and the small pistillate flowers on separate plants; the pistillate flowers contain abortive and useless stamens. *Calyx-teeth* nearly equal, semiovate. *Corolla-tube* longer than calyx in hermaphrodite, subequal in pistillate form, covered with glands. *Stamens* abortive in pistillate form (Fig. A 2); the two longer ones in the hermaphrodite forms exserted and divergent, the shorter pair included. *Stigmas* short, subequal. *Achene* ovate-oblong, compressed, pointed at base. *Leaves* entire, ovate, whitish or purple beneath, pubescent.

Origanum vulgare, Linn., β. *prismaticum*, Gaud. Fl. Helv. iv. p. 78; Gren. et Godr. Fl. de Fr. ii. 656.

(C, D.) Gen. Char.—*Flowers* in axillary cymes. *Calyx* tubercular during flowering (ventricose in C. alpina, Lam., and C. Acinos, Clairv.), 2-lipped, 13-nerved, throat hairy. *Corolla* 2-lipped, the upper lip somewhat concave, the lower of 3 broad lobes. *Stamens* (when perfect) convergent under the upper lip, the lower (outer) pair longest; *anther-cells* separated, divergent at base.

(C, D.) Spec. Char.—*Flowers* in dichotomous cymes; those of the large-flowered hermaphrodite form (C) and those of the small-flowered pistillate form (D) on separate plants; the cymes of flowers being much more crowded in the latter than in the former case. *Calyx* smaller in the pistillate form. *Corolla* smaller and less brightly coloured in the pistillate form. *Achene* subglobose, punctate, in the pistillate, ovoid-oblong in the hermaphrodite form. *Leaves* pubescent.

Calamintha Nepeta, Link et Hoffm. Fl. Port. p. 141; Gren. et Godr. Fl. de Fr. p. 664; Woods, Tour. Fl. p. 289.

Habitats.—All gathered by me at Mentone in November, 1867.

REMARKS.—Both Origanum vulgare, β. prismaticum, Gaud., and Calamintha Nepeta, Link, are common wayside plants in South France;
yet they give us an opportunity of studying a most curious and interesting instance of the partial separation of sexes which occurs in some
Labiatæ.   Calamintha Nepeta, Link, appears under two forms, the one
with large, brightly-coloured flowers, which contain both pistil and
stamens in a fully-developed and useful condition (Fig. 3); the other
with small, crowded flowers, in which the stamens are aborted and
quite useless, but the pistil is well formed (D). Now these hermaphrodite and pistillate plants are produced from seed of either one or
the other form, and, probably, from a given number of seeds, about half
of one kind and half of the other would grow. It is singular that the
achenes are unlike in shape, and that there should be so great a difference in the size of the flowers; for it is necessary that insects should
be attracted both to the hermaphrodite and the pistillate plants. However, the pistillate plant yields an immense quantity of seed,—more, I
believe, than the hermaphrodite. In Thymus vulgaris, Linn., and T.
Serpyllum, Linn., the plants are hermaphrodite and pistillate in like
manner, and the flowers differ similarly in size. Origanum vulgare,
β. prismaticum, Gaud., is a variety of our Wild Marjoram of England,
from which it is distinguished by its longer and less compact flowering
spikelets. This plant is also in the same condition with Calamintha
Nepeta, Link, as far as the flowers go. Now it is very interesting to
observe that plants may occasionally be found which have the abortive
stamens partly restored, and which are in an intermediate condition between the hermaphrodite and the pistillate forms, thus showing by reversion one of the steps through which the stamens were reduced to
their now useless state. This was well shown in some specimens of the
wild English Marjoram, observed by Mrs. Nevil Maskelyne at Dover.
One of these, of which only one plant was found, had "all the blossoms
on a small head of open flowers with 2 abortive stamens, 2 good long
ones and a long style." Another plant was in a very polymorphic condition, for it had, in the same head, "one blossom with 3 stamens longer
than the short style, and 1 abortive stamen;" three blossoms with 3
abortive and one perfect stamen; one in which all the 4 stamens were
aborted; and one in which 2 stamens were aborted and 2 perfect.
Now it is natural to ask what advantage the Calaminth, Thyme, and
Marjoram gain by having, in the one form, aborted stamens which are
not available for fertilization. The answer is, that foremost among the
laws which govern the reproduction of all living creatures, is one which
requires, at least occasionally, the union of two distinct individuals,
for perfect fertility and for the production of healthy offspring. Now,
though one might suppose that a flower in which the pollen and ovaries
are side by side would be sure to fertilize itself in the most effective
manner, yet this has been found not to be the rule. In a great many
cases the pollen is much better adapted for the fertilization of another

individual than it is for impregnating the ovules of its own flower. The advantage gained by fertilizing one plant with the pollen from a distinct individual of the same species is admirably demonstrated by some experiments recorded by Mr. Darwin,[*] in which he shows that seeds from self-fertilized and crossed (one individual by pollen of another plant of the same species) unions, when sowed close together and equally favoured, prove, by their growth, in their struggle for mastery, that the latter are stronger and more productive than the former. We have seen that the pistillate-flowered Calamintha produced a great abundance of seed, and we may now, I think, conclude that this species is all the more flourishing and vigorous because it produces two forms, thus necessitating the importation of pollen from a distinct individual.

EXPLANATION OF PLATE LXII.—Fig. A 1, pistillate flower. A 2, corolla of the same cut open. A 3, calyx cut open. A 4, an achene. A 5, the same, of the natural size ; all the Figs., except A 5, magnified. Fig. B 1, flower from a distinct, hermaphrodite plant, magnified. B 2, corolla of the same cut open. Fig. C 1, hermaphrodite flower, magnified. C 2, corolla of the same cut open, magnified. C 3, an achene, magnified. C 4, the same, of the natural size. Fig. D 1, pistillate flower, magnified. D 2, corolla of the same cut open, magnified. D 3, fruiting calyx, magnified. D 4, achene, magnified. D 5, the same, of the natural size.

[*] 'Animals and Plants under Domestication,' ii. 127.

A.

B.

C.

D.1. D.2.

C.2.

C.1.

PLATE LXIII.

(A, B, C, D) PRIMULA Allionii, Lois.

Natural Order PRIMULACEÆ.

GEN. CHAR.—See description of Plate XI. Part I.

SPEC. CHAR.—*Flowers* dimorphic, 1–2 (with sometimes an abortive third bud), on an extremely short peduncle. *Bracts* membranous, reniform or very broadly ovate, wrapped round the pedicels. *Calyx-lobes* rounded. *Corolla-lobes* emarginate. *Capsules* glabrous, deeply cleft into ovate, acute, upright lobes, nearly as long as the calyx. *Leaves* obovate or subspathulate, when fully developed narrowed into a petiole, slightly and irregularly toothed, densely covered with short, glandular pubescence, as are all parts of the plant except the corolla and the capsule.

Primula Allionii, Lois, Notice, p. 38, tab. 3, Fig. 1; Woods, Tour. Fl. p. 302, excluding the habitat.

HABITAT.—A, B, C, D, all collected by my father in the Gorge Sauvage of the Vallée de Caïros, near Saorge (Alpes Maritimes). A and B on April 1, 1868; C, April 26, 1867; D, April 23, 1864.

REMARKS.—It has been stated that Primula Allionii, Lois, grows among the Dolomite mountains of Tyrol, and especially in the southern part of that district. The Tyrol plant is, however, the distinct though closely related species Primula Tyrolensis, Schott,[*] separated at once from P. Allionii, Lois, by its linear or cuneate, herbaceous, divergent bracts; while those of our plant are reniform or broadly ovate, membranous, and overlapping each other. I have compared the specimens of the Tyrol plant in the Kew Herbarium (P. Tyrolensis, Schott, Monte Serva, Belluno, Papperitz, ex Herb. Fl. G. Rchb.) and those in the British Museum (2060, P. Allionii, Lois, ‘Tirol in Fiemme auf den Alpe Castellazzo di Paneveggio,’ Dr. Facchini) with the true Primula Allionii, Lois, which I figure, and find that they differ in the characters above mentioned, as well as in less important points. The same observations also apply to a specimen wrongly labelled P. Allionii, Lois, given me by Mr. G. C. Churchill, whose name is now so well known in connection with the Dolomite Mountains, gathered by Ambrosi (Val Caldiera, in Val Sugana, South Tyrol). Mr. Churchill, in a letter to me, speaking of this Tyrol Primula, says, "I gathered it on Monte Civita, south of Caprile, in the province of Belluno, and certainly on Dolomite." Primula Allionii, Lois, grows exclusively in the mountainous regions north of Mentone, and the only habitats yet found are

[*] Schott, Sippen. p. 13, ex Rchb. Ic. Fl. Germ. xvii. p. 44, tab. 1101, fig. 3.

the following :—the Madonna di Finestre, Entracque, a village in Piedmont, near Valdieri, and the Gorge Sauvage of the Vallée de Caïros, mentioned above.

EXPLANATION OF PLATE LXIII.—Fig. C 1, a fruiting calyx, peduncle, pedicel, and bracts, showing two abortive buds. C 2, the calyx of the same, divided longitudinally, showing the immature capsule. Fig. D 1, an inflorescence, with part of one corolla remaining, divided longitudinally. D 2, a bract. All the Figs. magnified.

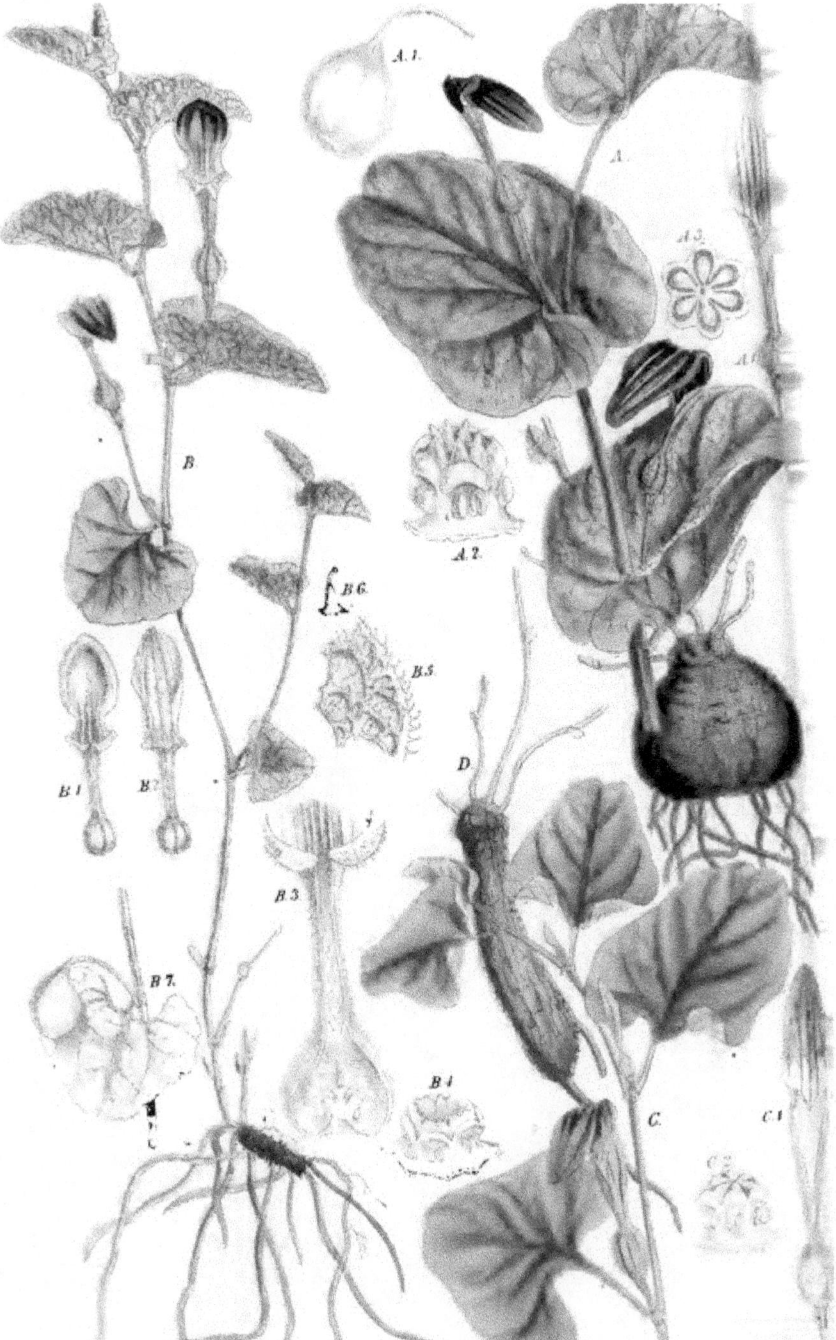

A.1.
A.
A.3.
A.2.
B.
B.6.
B.5.
B.1
D.
B.3.
B.7.
B.4
C.
C.1

PLATE LXIV.

(A) ARISTOLOCHIA ROTUNDA, Linn.!; (B) A. PISTOLOCHIA, Linn.!;
(C) A. LONGA, auctorum, non Linn.

Natural Order ARISTOLOCHIACEÆ.

GEN. CHAR. — *Calyx* or *perianth* coloured, tubular, swollen above the ovary; ... limb ... usually 1-2-lipped ... *Anthers* 6, rarely 5, extrorse, bilocular, ... completely united along the back to the column bearing the styles. *Ovary* inferior, 6-celled (5-celled in plants having only 5 anthers); ovules numerous, anatropal, horizontal, inserted in two rows. ... *Capsule* 6-celled, many-seeded, 6-valved (5-valved in pentandrous plants); dehiscence septicidal, rarely from the apex, usually from the base. ... *Seeds* many, ... often boat-shaped, with inflexed edges, upper face concave, the prominent, thickened raphe forming a longitudinal ridge in the centre; albumen fleshy, radicle turned towards the hilum; cotyledons 2, equal or nearly equal ... Duchartre in DC. Prodr. xv. part i. 432.

(A.) SPEC. CHAR.—*Flowers* greenish below, purple-brown within and above, solitary. *Peduncle* about an inch in length. *Perianth-tube* subcylindrical, about as long as the suboblong lip, swollen base globular. *Stigmatic lobes* conical. *Leaves* glabrous, orbicular-oval, cordate, basal lobes touching or overlapping one another; petiole extremely short. *Rootstock* subglobular.

Aristolochia rotunda, Linn.! Sp. Plant. p. 1316; Gren. et Godr. Fl. de Fr. iii. 73; Woods, Tour. Fl. p. 324.

(B.) SPEC. CHAR. — *Flowers* greenish below, purple-brown above, solitary; *peduncle* an inch or more long. *Perianth-tube* slightly shorter than the lip, pubescent, nearly cylindrical, but having the edges of the orifice dilated and recurved; *lip* concave, ovate, contracted below, *swollen base of tube* globular. *Stigmatic lobes* broader than long, emarginate, compressed, receding so as to leave a circular cavity in the centre. *Leaves* dull green, covered with short, harsh hairs, broadly cordate-ovate, emarginate, with a central mucro, basal lobes widely separated; *petiole* extremely short. *Rootstock* slender, cylindrical, irregularly articulate, emitting a dense mass of long, cylindrical fibres.

Aristolochia Pistolochia, Linn.! Sp. Plant. p. 1316; Gren. et Godr. Fl. de Fr. iii. 72; Woods, Tour. Fl. p. 324.

(C.) SPEC. CHAR.—*Flowers* greenish below and in the throat, brown above; *peduncles* very short, scarcely a quarter of an inch in length. *Perianth-tube* clavate, one-third longer than the oblong-ovate slightly attenuate lip; *swollen base of tube* oblong. *Stigmatic lobes* confluent, wedge-shaped, radiating from the centre. *Leaves* glabrous, broadly cordate-ovate, subemarginate, basal lobes widely separated ; *petiole* about an inch long. (*Rootstock* wanting in my specimens, but described by the discoverer (Rev. W. Hawker) and by M. Ardoino as being fusiform. This fusiform or nearly conical shape is that assumed by the rootstock when young, it afterwards becomes cylindrical.)

Aristolochia longa, auctorum, non Linn., ex Herb. *A. longa, β,* Linn. Sp. Plant. ed. 1, p. 962. *A. longa,* Gren. et Godr. (non Linn.) Fl. de Fr. iii. 73; Woods, Tour. Fl. p. 324; Duchartre in DC. Prodr. xvi. part ii. p. 487.

Habitats.—(A) Gathered by me at Mentone, April 2, 1867 ; (B) collected by the Rev. W. Hawker on Mont Mulacé, April 22, 1867 ; (C) gathered by me on the eastern slopes of Mont Coudon, near Hyères, May 7, 1868 ; (D) root and rootstock of a specimen labelled "Aristolochia longa, Linn., Matriti (Madrid) in umbrosis Real cam campo, April, 1841, Reuter," in Gay's Herbarium.

Remarks.—Aristolochia rotunda, Linn., abounds along the coast from Genoa to Marseilles. A. Pistolochia, Linn., has been found at Levens, and Gilletta, near Nice, and in the Esterelles mountains (Ardoino), and on the mountains near Hyères. A. pallida, Willd. (not figured), grows on the mountains near Genoa (Dutrs.), and was formerly gathered near Nice and at Torretta-Revest (Ardoino), and has lately been discovered for the first time in France (excepting the neighbourhood of Nice) on the Sainte-Baume, north of Toulon, by Dr. Shuttleworth ; this species has a globose root, but differs from A. rotunda, Linn., in its leaves, the petioles of which are half an inch long, and also in the flowers,* which have an extremely short lip, only one-fourth of the length of the tube, and are said to be pale yellow-green, with purplish veins, and a dark purple blotch at throat. A. longa, auct., grows in the olive-yards at Porto Mauritzio (Dutrs., who says that the lip in his specimens from this locality is not acute, but obtuse or retuse with a mucro), and on the Mulacé and Grammont mountains, near Mentone, where the Rev. W. Hawker detected it. Hitherto no doubt seems to have been entertained about the identity of the plant found about Montpelier, in the Pyrenees, etc., with the Aristolochia longa of Linnæus. But, on consulting Linnæus's Herbarium, I found that the plant below which Linnæus had written "12. longa," is a wholly different though closely allied species, having flowers (by measurement) 3 inches in length. This plant, which Linnæus named and numbered in accordance with A. longa, the twelfth species in the first edition of the 'Species Plantarum,' is certainly not the 'longa' of modern authors, but the fine, allied Algerian species described by Boissier and Reuter† as A. Fontanesii. There seems no possibility of any change or mistake in the specimen,

---

* Sibth. et Sm. in the 'Flora Græca,' x. tab. 936, represent the stigmatic lobes as purple, and united so as to form a central cone ; p. 28, "Columna alba, sexdentata, medio stigmate purpureo umbonata."　　　† Pugill. Pl. Nov. p. 108.

for Linnæus has repeated the number "12" on the strip of paper
which fastens the plant down. Therefore, A. Fontanesii should properly be called A. longa, Linn. Now, Sir J. E. Smith has written this
note under the Linnean specimen alluded to, "Vix II. B.   A. longa
hispanica, II. Jacq.," which probably means that a specimen in the
Banksian Herbarium does not correspond with that of Linnæus, and
the latter half may perhaps mean that the A. longa hispanica of Jacquin's Herbarium was the plant which he had supposed to be the true
longa. Jacquin's specimen in the Banksian Herbarium named by Solander "A. longa, β. (Jacquin)," is the small-flowered plant which I
have figured, and which is commonly but erroneously called A. longa.
This specimen is now in the Herbarium of the British Museum. It
is probable, therefore, that Jacquin understood that his plant was the
variety of β. hispanica, Linn. (Sp. Plant. ed. 1, p. 962), and not the
true, large-flowered A. longa, Linn. More observations on living specimens of these species of Aristolochia are wanting, and I would draw
the attention of botanists to the differences which the stigmatic lobes
present.

EXPLANATION OF PLATE LXIV.—Fig. A 1, a flower of the natural
size. A 2, the stamens and column magnified. A 3, a transverse section of the ovary, magnified. A 4, fruit of the natural size. Fig. B 1,
front, and B 2, back view of a flower of the natural size. B 3, the tube
of the perianth longitudinally divided, magnified. B 4, the stamens and
column magnified. B 5, portion of back of leaf, magnified. B 6, one
of the hairs. B 7, leaf and immature fruit of the natural size. Fig. C 1,
flower of the natural size. C 2, stamens and column, magnified. Fig. D,
rootstock of the natural size.

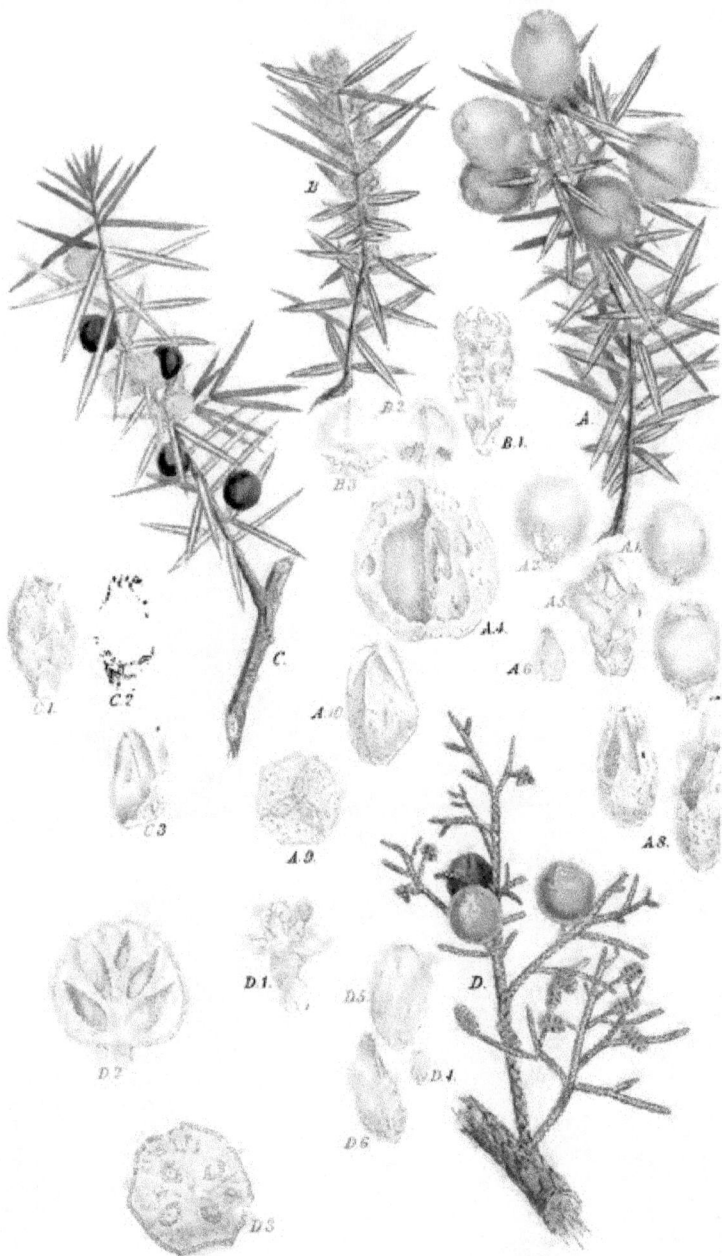

PLATE LXV.

(A, B) JUNIPERUS OXYCEDRUS, Linn., var. *macrocarpa*;
(C) J. COMMUNIS, Linn.; (D) J. PHŒNICEA, Linn.

Natural Order CONIFERÆ.

GEN. CHAR.—*Flowers* diœcious, or monœcious on separate branches. *Male catkins* axillary or terminating the lateral branches; *bracts* . . . bearing the stamens below on the lower side of their stalk; *filaments* wanting; *anthers* 3-6 under each bract . . . unilocular . . . *Female catkins* axillary or terminating the lateral branches, surrounded at the base by a few sterile bracts; *scales* 4-6, decussately opposite, or 3-9, and ternately verticillate, . . . composed* of the leafy bract and the fleshy, subequal, scale closely combined into one . . . *Ovary* (ovule of most authors) orbicular, sublenticular, or compressed. *Style* very short . . .; *stigma* almost orbicular . . . *Galbulus* ripening during the second year, composed of fleshy scales and . . . 1-3 or rarely 4-8 nuts. *Nuts* erect . . ., distinct, or very rarely united into a two- or three-celled nut . . ., pericarp stony. *Seed* solitary, . . .; *embryo* in the axis of fleshy albumen. Parl. in DC. Prodr. xvi. pars ii. p. 475.

(A, B.) SPEC. CHAR.—*Catkins* diœcious. *Galbuli* large, usually as long as or longer than leaf, covered with bluish-white bloom, when mature, red-brown, more or less bright, variable in shape, either globular or pyriform on the same branch, never prominently lobed. *Nuts* 3 only, ovoid-oblong, emarginate, with the persistent style forming an apiculum in the notch, having 2-3 prominent angles, partly enclosed in a bilobed membrane. *Leaves* linear-lanceolate, acute and pungent, the lower ones on each branch ovate or ovate-lanceolate, all having two whitish lines on upper surface, lower surface angular. *Growth* that of a small tree, attaining to about the size of a large Hawthorn (Cratægus Oxyacantha, Linn.).

Juniperus Oxycedrus, Linn., α. *macrocarpa*, Moggridge. *J. Oxycedrus*, Endlicher, Syn. Conif. p. 10. *J. macrocarpa*, Sibth. et Sm. ex Spec. in Herb. Sibth. non ex deser. in Prod. Fl. Græc. ii. 263. *J. macrocarpa*, Parl. in DC. Prodr. xvi. pars ii. p. 476.

(C.) SPEC. CHAR.—*Catkins* diœcious. *Galbuli* small, purplish-black when ripe. *Nuts* 3 only, ovoid-oblong, having 2-3 prominent angles, partly covered by an adherent membrane. *Leaves* linear, pungent, having only one whitish, central line. *Growth* shrubby, attaining 3-9 feet.

' * M. Parlatore regards each scale, at the base of which ovules are formed, as the bract of a lateral aborted branch, one or two other bracts belonging to which branch are united in part, or completely, as in this case, with the inner face of the scale.

Juniperus communis, Linn. Sp. Plant. p. 1470; Gren. et Godr. Fl. de Fr. iii. p. 157; Woods, Tour. Fl. p. 343.

(D.) Spec. Char. — *Catkins* monœcious (or sometimes diœcious, whole bushes only producing either male or female catkins). *Galbuli* globular or globular-depressed, red-brown when ripe, composed of 6, 8, 9, or rarely 10 scales. *Nuts* 6–9, oblong, compressed, apiculate. *Leaves* decussately arranged in pairs or in whorls of 3 each (both systems occur on the same branch). *Growth* low, shrubby, rarely exceeding 4 feet.

Juniperus phœnicea, Linn. Sp. Plant. p. 1471; Gren. et Godr. Fl. de Fr. iii. 159; Woods, Tour. Fl. p. 343.

Habitats.—(A, B) Mentone, December 30, 1867; (C) Mentone, December 30, 1867; (D) Mentone, January 14, 1868.

Remarks.—Juniperus Oxycedrus, Linn., is found along the coast between Marseilles and Genoa, under two principal forms, namely, the var. macrocarpa, here figured, which abounds about Nice and to the eastward, and the var. rufescens (J. rufescens, Link, distinguished by narrower, linear leaves and smaller galbuli), which is predominant from the Esterelles to the westward. About Cannes, among the Esterelles mountains, and at Fréjus, there is a form intermediate in all respects, which was pointed out to me by Dr. Shuttleworth under the name of var. intermedia (J. macrocarpa, Carrière). There is a great puzzle about J. macrocarpa, Sibth. et Sm., for the specimens preserved in Sibthorp's Herbarium at Oxford* do not answer to their description in the Prodromus Fl. Græc. ii. 263; the fruits of the specimen being of a dull red, and corresponding well with the plant figured here, and not of the black colour with the blue bloom (*nigræ cum rore cæruleo*) described. I have hunted through a great many collections, but have never found any variety presenting fruits of a dark blue colour. Dr. Shuttleworth kindly sent me specimens of a fine Juniper from Messrs. Huber's gardens at Hyères, which was remarkable for its very broad, dark, bluntish leaves, its orbicular, compressed nut, and especially for the trilobed fruit. This was recognized by M. Cosson as the true J. macrocarpa, Sibth. et Sm., with which he was familiar in Algeria. The fruit, however, was reddish. This is not the J. macrocarpa described by Parlatore,† who distinctly states that the scales forming the fruit do not project, but are closely combined, and that the species, as he conceives it, has a very wide range, one of the special habitats given being the neighbourhood of Nice! De Notaris mentions a form‡ which he distinguishes as var. ericoides (Rep. Fl. Lig. part ii. (1848), p. 403), remarkable for its erect, fastigiate branches, and leaves of a uniform pale green, rounded on the back. I have seen Spanish specimens of the same variety. Juniperus communis, Linn., is very easily known by its

---

* I am indebted to Mr. Baxter, of the Botanic Garden, Oxford, for a description of the Sibthorpian specimen.

† In DC. Prodr. xvi. pars ii. p. 476.   ‡ Growing at Sestri di Ponente.

small, purplish-black fruits, and the one white line down the leaf. The compact var. of J. communis, Linn., distinguished by some as a species, under the name of J. nana, Willd., is remarkable for its densely packed, short, curved leaves; but is only found in high Alpine situations in the Alpes Maritimes. Juniperus phœnicea, Linn., belongs to quite another section of the genus (Sect. Sabina, Spach), and is distinguished at once by its scale-like leaves; on abnormal branches one may, however, sometimes see long, linear leaves like those of J. Oxycedrus, Linn., in shape. A curious form, distinguished by Parlatore as var. monstrosa (Parl. in DC. Prodr. l. c.), is found among the mountains near Mentone. In this the nuts are pointed, and project through the fruit, a condition perhaps due to injury from insects. M. Parlatore, whose generic description I have copied above, takes a different view of Conifers from that now generally accepted. Many botanists treat Coniferæ as a great section of the vegetable kingdom, wholly unlike other flowering plants in having the ovules and seeds without any covering, style, or pericarp. M. Parlatore believes that the ovule and seeds are enclosed in an ovary, in the tubular apex of which he sees a style and a stigma in its orifice. This view appears to me tenable, and has the attraction of making one mystery the less.

EXPLANATION OF PLATE LXV.—Fig. A 1, a subglobular fruit, in which only 6 scales are combined, of natural size. A 2, a subglobular fruit of the natural size, in which 3 scales from the third whorl from the apex are partly combined. A 3, a subturbinate fruit of the natural size, having these 3 scales combined. A 4, a fruit longitudinally divided, with only one nut remaining, showing the resinous cavities, magnified. A 5, female catkin, with 2 of the 3 scales of the uppermost whorl, and one ovary removed, magnified. A 6, an ovary, magnified. A 7, nut of the natural size. A 8, two views of the same, magnified, showing the bilobed, membranous coat. A 10, the same, with the membrane removed. A 9, an immature fruit, divided transversely in half, of the natural size. A 11, front view of leaf, magnified. A 12, transverse section of leaf, the back uppermost, magnified. Fig. B 1, male catkin, magnified. B 2, front view, and B 3, back view of stamen-bract. Fig. C 1, female catkin, magnified. C 2, the same, deprived of its scales, except the 3 inner ones, which differ in shape and texture, magnified. C 3, a nut, with its membranous coat, magnified. Fig. D 1, female catkin, magnified, showing the expanded scales and the ovaries. D 2, fruit magnified, showing 5 of their nuts or their cavities. D 3, transverse section of fruit, magnified. D 4, a nut of natural size. D 5 and D 6, different views of the same, magnified.

PLATE LXVI.

FRITILLARIA MONTANA, Hoppe.

Natural Order LILIACEÆ.

GEN. CHAR.—See description of Plate XXV. Part. I.

SPEC. CHAR. — *Flower* small, yellowish, densely chequered with purple-brown. *Divisions of perianth* oblong-elliptic, rounded at apex, the nectariferous depression at base, narrow-oblong. *Leaves* distant from one another, nearly straight, linear-oblong, the uppermost two or three forming a whorl rather distant from the flower, the lowest pair opposite or nearly so, the intermediate leaves alternate. *Stem* stiff, and nearly straight.

Fritillaria montana, Hoppe, Bot. Zeit. xv. pars 2, p. 476; Woods, Tour. Fl. p. 364. *Fritillaria caussolensis*, Goaty et Pons in Ardoino, Fl. Alp. Mar. p. 375. (OBS. The specimens from which this description and drawing were made were not absolutely fresh, having been in press for some hours before I received them. For these I am indebted to the kindness of M. Thuret.)

HABITAT.—Caussols, near Grasse (Alpes Maritimes), collected by M. Huet on April 22, 1867.

REMARKS.—This plant is nearly allied to Fritillaria involucrata, All., and is principally distinguished by its smaller and darker flowers, and stiffer, more scattered leaves. The European members of this genus are not separated from one another by well-marked characters, the differences between them being chiefly those of aspect and habit, such, in fact, as gardeners and not botanists are wont to make use of. However, no one who had before him fifty or sixty living specimens of F. involucrata, All., mixed with an equal number of specimens of F. montana, Hoppe, would have any difficulty in separating the one from the other. The recent re-discovery of F. montana, Hoppe, near Grasse, is one of great interest, as the plant has hitherto been only cited as found in Istria. I say re-discovery, for there are specimens in the Herbarium of the late M. Gay, at Kew, labelled, " F. meleagris, Caussols (Var.), Hassenot misit; Perreymond, April, 1838." M. Gay has added below, " F. montana, Hoppe ?"

EXPLANATION OF PLATE LXVI.—Fig. 1, an inner and outer division of the perianth, viewed from within, showing the nectariferous hollow; of the natural size.

A. *A.8* *A.1* *A.2* *A.7* *A.6*

B. *B.2* *B.1* *A.5* *A.3* *A.4*

PLATE LXVII.

(A, B.) ORNITHOGALUM comosum, Linn.

Natural Order LILIACEÆ.

GEN. CHAR.—*Spathe* o. *Peduncles* not jointed. *Perianth* spreading, divisions distinct, without nectaries. *Stamens* almost free from the divisions of the perianth; *filaments* inserted at the backs of the anther, about midway up. *Style* entire.

SPEC. CHAR.—*Flowers* at first in a compact corymb, lengthened out, when in fruit, into an oblong raceme; *fruiting peduncles* ascending; *bracts* large, spathe-like, enveloping and exceeding the peduncles. *Outer divisions of perianth* oblong-lanceolate, obtuse, with mucro, and hooded at apex; *inner divisions* broadly-lanceolate. *Stamens* bent outwards after dehiscence; *filaments* tapering from below the middle upwards, filiform above. *Leaves* 3–6, broad, channelled, tapering from the base upwards, having an indistinct central line. *Bulb* simple, usually subglobose.

Ornithogalum comosum, Linn. Sp. Plant. p. 440; Rchb. Ic. Fl. Germ. x. (1848), tab. 468, fig. 1021, p. 15; Woods, Tour. Fl. p. 368.

HABITAT.—(A, B) Porta degli Angeli, Genoa, where I gathered it, April 16, 1868.

REMARKS.—This very fine Ornithogalum is easily distinguishable from O. divergens, Bor., so common along the coast, by its single bulb without offsets, stiff, broad leaves, and peduncles much longer than the bracts. It is also a much more compact and handsome plant. When Mr. G. C. Churchill was my companion on a botanical excursion last April, we had the good fortune to discover this handsome species growing on the grassy slopes below the Porta degli Angeli, at Genoa. Ornithogalum comosum, Linn., has not been recorded before from any habitat along the Riviera. According to Parlatore, this species has been found in Istria, in Northern Italy, on Monte Fortino, close to Valle, a town on the river Po, not far from Mortara, in North-eastern Piedmont; in Southern Italy, on Monte Velino, near Pizzoli, in the province of Abruzzo, on Monte Gargano, close to Spigno, etc. The general range given by Nyman comprises habitats in Austria, Dalmatia, Croatia, Hungary, Transylvania and Macedonia. There are specimens of O. comosum, Linn., in the Herbarium of M. Gay, collected by Pauer and Sadler, near Buda.

EXPLANATION OF PLATE LXVII. — Fig. A 1, A 2, an outer and inner division of perianth of the natural size. A 3, apex of outer division, magnified. A 4, A 5, stamens of the natural size. A 6, A 7, the same, magnified. A 8, portion of a leaf of the natural size. Fig. B 1, a seed of the natural size. B 2, the same, magnified.

PLATE LXVIII.

(A, B.) ORNITHOGALUM EXSCAPUM, Ten.

Natural Order LILIACEÆ.

GEN. CHAR.—See description of preceding Plate.

SPEC. CHAR.—*Flowers* pale. *Fruiting peduncles* swollen at base, bent back so as to be parallel to the very short scape, with the fruit ascending. *Bracts* (in specimens) ovate-oblong, pellucid, only about half as long as peduncles. *Filaments of stamens* tapering from the base upwards, filiform above. *Capsule* broadly-ovate. *Leaves* slightly channelled, with narrow, white line. *Bulb* simple, ovate.

Ornithogalum exscapum, Ten. Fl. Nap. i. 175; Parl. Fl. ii. 427; Woods, Tour. Fl. p. 368. *O. mutabile*, Dntrs. Rep. Fl. Lig. pars ii. p. 440.

HABITAT.—Gathered by me in the valley tributary to the Varena torrent at Pegli, near Genoa, April 17, 1868.

REMARKS.—Ornithogalum exscapum, Ten., is remarkable for its reflexed fruiting peduncles, the extreme shortness of the scape, the flowers appearing to spring from the ground, and for its single bulb; this latter character separates it from O. refractum, Kit.,[*] to which it is otherwise most closely allied. The characters drawn from the bulbs in Ornithogalum and Gagea appear to me to hold good in the great majority of species when wild. I gathered a large series of specimens of O. exscapum, Ten., but in no case was there any bulblet attached to the bulb, though I found one bulb which produced two fine flower-scapes. Parlatore mentions this species in the neighbourhood of Genoa, in Sicily, Corsica, and Sardinia, but it appears to be a scarce plant. There is, however, no description of O. refractum, Kit., given in the ‘Flora Italiana,’ and we may conclude that it is not known in Italy.

EXPLANATION OF PLATE LXVIII.—Figs. A 1, A 2, an outer and inner division of the perianth. Figs. A 3, A 4, two stamens of the natural size. Figs. A 5, A 6, the same, magnified. A 7, part of a leaf, magnified. A 8, a capsule, nearly mature, of the natural size. A 9, the apex of the same. Fig. B, a fruiting raceme of the natural size.

---

[*] See Schultes, Syst. Veg. vii. 532. This species has been confused by most authors with O. exscapum, Ten., the character in the bulb having been overlooked.

PLATE LXIX.

(A, B) ASPHODELUS cerasiferus, Gay; (C) A. albus, Mill.;
(D) A. microcarpus, Viv.

Natural Order LILIACEÆ.

GEN. CHAR.—*Flowers* yellow white or pinkish. *Pedicels* solitary,
or sometimes fasciculate in the axils of the bracts, jointed. *Divisions of
the perianth* spreading. *Stamens* having their filaments dilated at the
base, so as to enclose the ovary. *Style* entire. *Seeds* angular. " Plants,
annual or perennial, or only flowering once after having lived many
years under the form of a simple rosette. *Radical fibres* very slender
in the first case (when annual); cylindrical, and of medium size in the
third (when a rosette, only flowering once); or fusiform, fleshy, and
sometimes very thick, in the second case (when perennial), and then
serving to nourish one or several axillary buds on the neck of the root
(*collet*), by means of which the plant perpetuates itself after the decay
of the stem, which is always annual, though the root is perennial."*

(A, B.) SPEC. CHAR.—*Flowers* large; bracts at first blackish (generally fulvous, Gay). *Divisions of perianth* rounded at apex; nerve
russet-coloured (nerve flesh-coloured, Gay). *Filaments* longer than
the divisions of perianth, protruded from the buds, the claws abruptly
contracted above, slightly papillose at edge only. *Fruit* turning yellow,
globular, depressed, as large as a cherry, when ripe (in specimen B)
having the valves transversely furrowed, but without prominent veins.
Stem unbranched, or having one or two short, simple branches. *Radical fibres* swollen, the enlarged part being nearly cylindrical, narrow,
about eight times as long as broad, gradually thickened downwards.

Asphodelus cerasiferus, Gay, Trois. Esp. Asphod. in Ann. Sc. Nat.
4me. sér. tom. vii. cah. 2. *A. albus*, Gren. et Godr. (non Willd.) Fl. de
Fr. iii. 224.

(C.) SPEC. CHAR.—" Bracts brown-black. *Divisions of perianth* having a green, central nerve. *Filaments* papillose-scabrous up to the
middle; the claws oblong, cuncate-ovate, gradually attenuated above.
Capsule of intermediate side, ellipsoid. *Stem* simple or shortly branched."
Gay, Ann. Sc. l. c.

Asphodelus albus, Mill. Gard. Dict. (8vo ed. 1768) i. 3; Willd. Sp.
Plant. (1799) p. 133. *A. sphærocarpus*, Gren. et Godr. Fl. de Fr. iii.
223, et *A. subalpinus*, Gren. et Godr. l. c. p. 224.

(D.) SPEC. CHAR.—*Divisions of perianth* oblong-lanceolate, scarcely

---

* The part within inverted commas is translated from an unpublished monograph
of Asphodelus by M. J. Gay (1856 to 1860).

shorter than stamens; nerve flesh-coloured. *Claw of filaments* obovate, densely papillose at margin. Fruit very small. *Stem* much branched. *Radical fibres* swollen, the enlarged part obovate, only about twice as long as broad.

Asphodelus microcarpus, Viv. excl. syn. Boiss. Fl. Cors. p. 5; Gren. et Godr. Fl. de Fr. iii. 223. *A. ramosus*, Woods, Tour. Fl. p. 365.

HABITATS.—(A.) Gathered by me on Mont Coudon, near Hyères, May 7, 1868. (B.) Capsule of Asphodelus cerasiferus, Gay, gathered by my father, March 12, 1867, on Mont Mulacé, Mentone. (C.) Capsule of Asphodelus albus, Mill., from a specimen in M. Gay's Herbarium, from Napoléon Vendée, collected by M. Pontarlier, July 5, 1856. (D.) Collected near Monaco by my father; flowers, March 20, and fruit, January 10, 1867.

REMARKS.—The three species of Asphodel here alluded to, form the subject of an excellent paper by M. Gay in the 'Annales des Sciences' quoted above. They are, as he tells us, not only distinguished by the characters already cited, but also by the districts which they inhabit. The small-fruited Asphodel (A. microcarpus, Viv.) loves the hot shores of the Mediterranean and its islands; the White Asphodel (A. albus, Mill.) usually chooses high alpine situations, or descends into the cooler plain-country of Western France; while the Cherry Asphodel (A. cerasiferus, Gay) prefers the mountains near the Mediterranean, where it can always be within sight of the olive-trees. I shall now give an abstract of the principal stations mentioned by M. Gay, either in his paper alluded to above or in a later, unpublished manuscript of 1856. When quoting from the latter, I shall place MSS. after the locality. Asphodelus microcarpus, Viv., is found in the Canary Islands, in Teneriffe, Palma, Lancerota (MSS.); Estremadura and Southern Portugal (MSS.), Cadiz, in Seville (MSS.), in Algeria, Tunis, Egypt, and Syria, on the shores of Asia Minor and of the Sea of Marmora, in the Ile des Princes, near Constantinople (Thuret (MSS.), Zante (MSS.), abundantly in Greece (MSS.), Fiume, and at Rome (MSS.). Along the Riviera, and as far as Marseilles, I only know of A. microcarpus, Viv., as being found in very small quantity near Monaco, abundantly on the Ile Sainte Marguerite, at Auribeau (Ardoino) and Agay, near Cannes, and in profusion about Hyères. A. albus, Mill., is widely spread in the Sierra di Guadamarra, in the chain of the Alps, the Apennines, and even on the further side of the Adriatic, but not reaching into Hungary, Servia, or Roumelia; amongst these mountains it ranges from the zone of the beech-tree to a maximum of 6500 feet; it can grow near the sea-shore, as for example at Trieste; in the neighbourhood of mountains at Nettuno, near Rome, and especially in the south-west of France, where it reaches its extreme northern limit in latitude 49°. From Marseilles to Genoa, Asphodelus albus, Mill., seems to be not uncommon in the mountainous and alpine region. Asphodelus cerasiferus, Gay, is wanting in the Canaries, and

very rare in Algeria, but is found plentifully in Granada and Andalusia, attaining 6000 feet in the Sierra Nevada, in the Sierra de Fuan
Santa, near Murcia; and at Collioure in the Pyrénées Orientales, in
Languedoc, south of Nîmes, Montpellier and Beziers; on the narrow
strip of land between Cette and Agde, which separates the Mediterranean from the Étang de Thau; on the Pic St. Loup, at Baume Oriol;
in the "Quartier de Touris" (where the specimen figured was gathered),
near Toulon; on Monte Niolo, Monte Cagna, near Talbuccio, near the
road between Bastia and Corte, and in the Forest of Aitone, between
Vico and Calvi, in Corsica (MSS.). In the district referred to in this
work, the known stations for A. cerasiferus, Gay, are those in the
neighbourhood of Marseilles (Montredon, etc.), and of Toulon and
Hyères, in the Esterelles, and on Mont Mulacé, near Mentone. It
probably is found between Mentone and Genoa, but accurate observations are wanting. M. Gay also acknowledges the existence of a rare
intermediate form, which he distinguishes as A. cerasiferus, β. intermedius; this approaches towards A. albus, Mill., in having fruits rather
smaller, which do not turn orange, and black bracts. This variety,
β. intermedius, is synonymous with A. Villarsii, Verlot, and is found
on Mount Rachet, near Grenoble; on a hill near Turin (MSS.), and on
Monte Senario, near Florence (MSS.).

EXPLANATION OF PLATE LXIX.—Fig. A 1, two divisions of the
perianth, and two stamens, of the natural size. A 2, A 3, claw of a
longer and shorter stamen, magnified. Fig. B 1, capsule of A. cerasiferus, Gay, of the natural size. Fig. C, capsule of A. albus, Mill., of
the natural size. Fig. D 1, two divisions of the perianth, and two
stamens, of the natural size. D 2, one of the longer stamens, magnified.
D 3, part of a fruiting raceme, of the natural size. D 4, D 5, a seed,
magnified and of the natural size.

B.2. B.1.

B

A.2

A.1

A.3

PLATE LXX.

(A) NARCISSUS papyraceus, Gawl.; (B) N. papyraceus, β. incurvata.

Natural Order AMARYLLIDACEÆ.

GEN. CHAR.—See description of Plate XXII. Part I.

(A.) SPEC. CHAR.—*Flowers* entirely white (except the tube, which is usually greenish), numerous, in a loose umbel. *Scape* compressed, sharply two-edged. *Divisions of perianth* ovate-oblong. *Crown* sub-entire, sides wavy, upright, about one-quarter of the length of the divisions. *Leaves* dark glaucous-green, slightly channelled. *Bulb* large. *Narcissus papyraceus*, Gawl. Bot. Mag. xxiv. tab. 947; Parl. Fl. Ital. iii. 125. *N. niveus*, Lois, Narciss. p. 37; Gren. et Godr. Fl. de Fr. iii. 260; Woods, Tour. Fl. p. 361.

(B.) SPEC. CHAR. — *Flowers* entirely white, tube greenish, about 6 in a loose umbel. *Divisions of perianth* short, the outer ones ovate, the inner ovate-oblong, all curved inwards. *Crown* very short, entire, sides upright. *Leaves* and *scape* of N. papyraceus, Gawl.

Narcissus papyraceus, Gawl. β. *incurvata*, Moggridge.

HABITATS.—(A) Mentone, December 12, 1866; (B) Turin valley, Mentone, February 26, 1868; both gathered by myself.

REMARKS.—The species and the variety figured differ not only in the characters above cited, but also in their time of flowering; for Narcissus papyraceus, Gawl., may be gathered at Christmas-time about Mentone, while the var. *incurvata* is not out before February. The different species and varieties of Narcissus are very complex, and, as most of them are found chiefly in cultivated ground, any new variety is readily propagated by division of the bulbs, which takes place when the earth is broken up. I have seen living specimens of Narcissus papyraceus, Gawl., from Pegli, near Genoa (sent to me by Mrs. Tebbs), and from San Remo, where it is found in great profusion. M. De Notaris states that it grows about Sestri di Ponente.

EXPLANATION OF PLATE LXX.—Fig. A 1, a flower divided longitudinally in half. A 2, a transverse section of the leaf. A 3, a transverse section of the scape. Fig. B 1, a transverse section of the scape. B 2, a transverse section of a leaf. B 3, an inner, and B 4, an outer division of the perianth. All the Figures are of the natural size.

A.

A.1.

A.2.

A.3.

A.4.

B.

C.

C.1.

C.2.

C.3.

C.5.

C.6.

D.

D.2.

D.1.

PLATE LXXI.

(A) NARCISSUS PANIZZIANUS, Parl. ; (B, C, D) N. DUBIUS.
Gouan.

Natural Order AMARYLLIDACEÆ.

GEN. CHAR.—See description of Plate XXII. Part I.

(A.) SPEC. CHAR.—*Flowers* small, white, with a white tube, in a rather compact umbel. *Scape* very much compressed, and acutely two-edged. *Divisions of perianth* acute. *Crown* entire, sides somewhat curved, and the margin contracted, one-third of the divisions in length. *Tube* nearly twice as long as divisions of perianth. *Leaves* of a rather bright green, channelled.

Narcissus Panizzianus, Parl. Fl. Ital. iii. 128.

(B, C, D.) SPEC. CHAR.—*Flowers* small, nearly pure white when fully blown, 3-5, all sloping in the same direction and at the same angle. *Scape* subcylindrical, compressed, rush-like. *Divisions of perianth* very short, the outer suborbicular, the inner ovate. *Crown* more or less distinctly 3-lobed (in specimens C) or entire (specimen B), half as long as divisions of the perianth ; *tube* twice as long as the divisions. *Leaves* not channelled, semicylindrical below, flat above on the upper and curved on the under face. *Bulb* small.

Narcissus dubius, Gouan, Illus. 22 ; Gren. et Godr. Fl. de Fr. iii. 260 ; Woods, Tour. Fl. p. 360.

HABITATS.—(A.) San Remo, collected by Mr. F. E. Hubbard, and verified by M. Panizzi, February 7, 1867. (B.) From a specimen grown in M. Thuret's gardens at Antibes, gathered March 9, 1868. (C.) Specimens with bulbs, collected by Dr. Shuttleworth on Mont Coudon, near Hyères, March 24, 1868. (D.) Fruit collected by me in the same locality.

REMARKS.—Narcissus Panizzianus, Parl., appears to me a doubtful and yet distinguishable species. Its much smaller flowers and brighter green leaves give it a very distinct appearance from Narcissus papyraceus, Gawl., but it approaches much more closely to N. polyanthus, Lois., from which its fewer flowers of acute divisions and sharply two-edged flattened scape may perhaps separate it. I have never seen a wild specimen of L. polyanthus, Lois., and prefer, therefore, to reserve my opinion till I have more material to judge from. N. Panizzianus, Parl., is only known to grow at San Remo. Narcissus dubius, Gouan, is one of the grateful exceptions to the rule among Narcissi, for it is found in wild rocky situations in the mountains near Toulon and Hyères, and never, as far as I know, in cultivated ground. It will be observed that in the specimens from the garden at Antibes, the crown of the corolla is entire, while it is lobed in those from Mont Coudon; I find that specimens of this plant in the Herbarium of M. J. Gay exhibit similar varieties. Thus a specimen from near Marseilles has a lobed crown, while those from Aix, in Provence, and from Villeneuve-les-Avignon (Gard) have the crown entire. The fruits vary in the same umbel from pyriform-oblong to pyriform. I have observed that the flower which opens last often remains but half developed in Narcissus, having the characters and colour of a lately expanded bud. It is possible that the smaller and more pyriform fruits may be the product of these blossoms. I would suggest to experimentalists that a series of observations noting the characters of seedlings from the earliest and the latest flowers of individual plants, might lead to interesting and valuable results. To make the experiments complete, both the first and the last flowers should be fertilized with pollen from a distinct individual of the same species, and should be protected from insect agency. N. dubius, Gouan, is not found in Italy or along the Riviera*; but from Mont Coudon to the westward, at Marseilles, Avignon, Pont du Gard, and Bione, in Hérault (Gren. et Godr.), Aix, in Provence (Herb. Gay!), in Corsica, at Ajaccio (Bourgeau!), near Mont Cada about twenty-one miles north of Barcelona (Bourgeau! in Herb. Gay; this label and that of the preceding specimen being written by M. Gay), Pyrenees, and Greece (Nyman).

EXPLANATION OF PLATE LXXI.—Fig. A 1, a flower cut in half, longitudinally, of the natural size. A 2, a transverse section of the scape, of the natural size. A 3, transverse section of a leaf, of the natural size. A 4, the same, magnified. Fig. B, a flower from the Antibes, cultivated specimen, of the natural size. Fig. C 1, a transverse section of the scape, of the natural size. C 2, a transverse section of the upper, and C 3, of the lower part of a leaf, of the natural size. C 4, an outer, and C 5, an inner division of the perianth. C 6, a flower cut in half, longitudinally, of the natural size.

* Nyman, in his 'Sylloge,' mentions it as growing at Nice, but does not give his authority. This is a probably a mistake.

A.

A. 1.

A. 2.

B.

B. 1.

B. 2.

B. 3.

C.

C. 1.

D.

D. 2.

PLATE LXXII.

(A) OPHRYS SPECULUM, Link; (B) O. BOMBYLIFERA, Link; (C) O. INSECTIFERA, Linn., var. *Bertolonii;* (D) O. INSECTIFERA, Linn., var. *Philippi.*

Natural Order ORCHIDACEÆ.

GEN. CHAR—See description of Plate XLIII. Part II.

(A) SPEC. CHAR.—*Sepals* green, striped with purplish-brown, the uppermost forming a hood over the anther. *Petals* pink-brown, narrow-triangular, velvety. *Lip* brown, with yellow inside, and the obcordate-central spot shining, *(lapis lazuli)* blue, soon fading, densely hairy towards the edges, trilobed, the central lobe recurved, the lateral lobes triangular, directed forwards. *Anther* obtuse, bent backwards. *Stigmatic cavity* prolonged into two parallel shining walls, with 4 glands, two on either side. *Leaves* broadly oblong-lanceolate.

Ophrys speculum, Link in Schrad. Journ. Bot. (1799), ii. 324; Parl. Fl. It. iii. 555; Woods, Tour. Fl. p. 354.

(B.) SPEC. CHAR.—*Sepals* green. *Petals* pubescent, short, oblong, recurved. *Lip* deeply trilobed, central lobe subglobular, cleft horizontally at the apex, so that there is a cavity between the lower fold, which bears the short process, and the upper surface of the lip; *lateral lobes* conical, prolonged below. *Anther* red, obtuse, bent backwards. *Stigmatic cavity* enlarged into a deep hollow, the glabrous walls projecting, but deeply divided by a notch opposite to the anther. *Leaves* small, oblong-lanceolate. *Tubers* 3 or more (one broken off in specimen figured), one or more on stalks, 1–2 feet long.

Ophrys bombylifera, Link in Schrad. Journ. Bot. (1799) ii. 325; Gren. et Godr. Fl. de Fr. iii. 303; Woods, Tour. Fl. p. 353.

(C.) SPEC. CHAR.—*Sepals* white or pink. *Petals* pink, glabrous. *Lip* oblong (in specimens sometimes almost orbicular!), apiculate, lateral lobes conical, flattened (often absent), marking shield-shaped, distinct, and far from base, either with (as in the individual figured) or without a marginal line, depressed below the general surface of the labellum.

Ophrys insectifera, Linn., var. Bertolonii, Moggridge. *O. Bertolonii,* Moretti, Fl. Ital. Dec. Sext. p. 9; Woods, Tour. Fl. p. 354.

(D.) SPEC. CHAR.—*Sepals* white (or pink), ovate. *Petals* pubescent, linear, recurved. *Lip* suboblong, having a broad apiculum, lateral lobes very small, conical, placed at some distance down the lip and away from the base, *markings* complex, of yellow lines enclosing purplish, glabrous spaces.

Ophrys insectifera, Linn., var. Philippi, Moggridge. *O. Philippi,* Gren. Orch. de Toulon, Mem. Soc. Em. Doubs (1859), p. 11.

HABITATS.—(A) Eastern bay, Mentone, where I found one plant, April 10, 1866. (B) Gathered near Ventimiglia by my father, April 5, 1866. (C) Gathered by me at Mentone, April 15, 1866. (D) Originally from Toulon, now cultivated at Antibes by M. Thuret, to whose kindness I am indebted for this specimen, the only living one I have ever seen.

REMARKS.—Ophrys speculum, Link, is a very remarkable species, both on account of the structure of the stigmatic chamber and of the colour of the lip. I have never been able to find more than one plant, and this is the only recorded case of its discovery along the Riviera or in France. Parlatore (l. c.) says that it grows in Sicily (Palermo, etc.), Calabria, Sardinia, Algeria, Portugal, Spain, Greece, Rhodes, and Smyrna. Ophrys bombylifera, Link, has a very curious enlargement of the stigmatic cavity, so arranged that there is an entrance for insects exactly opposite to the anther, the pollen of which they are wanted to remove. This pretty little plant has been found at Mentone, in the Gorbio valley, by the Rev. W. Hawker; it grows also at Ventimiglia, La Brague, near Antibes, and at Toulon. Ophrys insectifera, var. Bertolonii, is a well-marked form, and easily distinguishable in the great majority of instances. Yet it is closely linked on by intermediates to the forms of var. aranifera, with pink sepals and purplish lips, so much so that many botanists take these intermediates for the types of what they consider a species. The var. Philippi is a very remarkable form, but I am assured by excellent observers that it is united by a multitude of intermediates to the var. arachnites (O. arachnites, Reich.). I have only seen one living specimen, and this I owe to the great kindness of of M. Thuret. I have diligently searched for var. Philippi, but I believe that it is only to be found in the neighbourhood of Toulon.

EXPLANATION OF PLATE LXXII.—Fig. A 1, anther, stigmatic cavity, and one petal. A 2, lip viewed from below. Fig. B 1, a flower deprived of the sepals. B 2, the same, viewed sideways. B 3, back view of a petal. Fig. C 1, lip. C 2, under side of lip. Fig. D 1, a petal. D 2, a flower deprived of sepals and petals. D 3, the lip, viewed from below. All the figures magnified.

PLATE LXXIII.

(A) ANDROPOGON DISTACHYOS, Linn.; (B) A. GRYLLUS, Linn.; (C) A. HIRTUM, Linn.; (D) A. PUBESCENS, Vis.; (E) A. ISCHÆ-MUM, Linn.

Natural Order GRAMINEÆ.

GEN. CHAR.—*Spikes* either solitary or united or digitate or disposed in a simple or compound panicle. *Spikelets* . . . in pairs (or, rarely, the terminal ones in threes), one pedicellate and male (in the European species), or female, neuter or abortive, the other sessile, 2-flowered, both or only one of the flowers being two-valved, the *lower flower* neuter or male (neuter, of one flowering glume only in the European species), the *upper* hermaphrodite or female. *Outer glumes* 2, becoming coriaceous, without or rarely with awns. *Flowering glumes* hyaline, the lower awned or not. *Stamens* 3. *Ovary* glabrous. *Styles* 2, terminal. *Stigmas* feathery. *Lodicules* 2, truncate. *Grain* glabrous, enfolded in the flowering outer glumes. Steudel, Syn. Glum. p. 363.

(A.) SPEC. CHAR.—*Spikelets* arranged in pairs, forming two ter-minal spikes, glabrous. *Sessile spikelets* 2-flowered; *lower outer glume* many-nerved, bifid above, coriaceous, glabrous, having 4 membranous edges, *upper outer glumes* membranous, 3-nerved, having an awn in the short notch at the apex rather longer than itself; *flowering glume of aborted flower* membranous, exaristate (Fig. A 3); *lower flowering glume of hermaphrodite flower* ovate-acuminate, deeply bifid, with an awn four to six times as long as itself (the *upper flowering glume* has been want-ing in all the specimens which I have dissected); *lodicules* 2, fleshy, sub-triangular. *Culms* unbranched.

Andropogon distachyos, Linn. Sp. Plant. p. 1481; Gren. et Godr. Fl. de Fr. iii. 467; Woods, Tour. Fl. p. 395.

(B.) SPEC. CHAR.—*Spikelets* 3, terminal, two pedicellate and male, one sessile, female (in specimens, usually said to be hermaphrodite); below these three there is often a pair of spikelets, one sessile and female, the other, pedicellate and male. *Sessile spikelet* 2-flowered, *lower outer glume* exaristate, coriaceous, the two prominent lateral nerves covered with short spines, *upper outer glume* awned, awn rather longer than itself, slightly spiny along dorsal nerve; *flowering glume of aborted flower* (B 4) pubescent, involute, exaristate; *lower flowering glume of female flower* linear, tapering into a long awn; *upper flowering glume of female flower* membranous, exaristate; *lodicules* 2, fleshy, sub-triangular. *Culm* bearing several whorls of filiform flower-branches.

Andropogon Gryllus, Linn. Sp. Plant. p. 1480; Gren. et Godr. Fl. de Fr. iii. 468; Woods, Tour. Fl. p. 395.

(C.) Spec. Char.—*Spikelets* in pairs along a hairy axis, forming a pair of spikes, with long hairs below the point of junction. *Sessile spikelet* 2-flowered, *lower outer glume* many-nerved, coriaceous, hairy on back, *upper outer glume* membranous, hairy above; *flowering glume of aborted flower* (C 4) membranous, oblong-lanceolate; *lower flowering glume of hermaphrodite flower* membranous, bifid, with long awn; *upper flowering glume* (C 5) minute, ovate, membranous, exaristate, about as long as the lodicules. *Culm* bearing many flower-branches, each sheathed in a leaf.

Andropogon hirtum, Linn. Sp. Plant. p. 1482; Gren. et Godr. Fl. de Fr. iii. 469; Woods, Tour. Fl. p. 395.

(D.) Spec. Char.—Differs from Andropogon hirtum, Linn., in its longer awns and spikes, and in having the stem glabrous below each pair of spikes.

Andropogon pubescens, Vis. Pl. Rar. Dalm. p. 3; Gren. et Godr. Fl. de Fr. iii. 469; Woods, Tour. Fl. p. 395.

(E.) Spec. Char.—*Spikelets* in pairs, forming a digitate terminal panicle of 4-8 spikes. *Sessile spikelet* 2-flowered, *lower outer glume* many-nerved, coriaceous, exaristate, *upper outer glume* membranous, exaristate; *flowering glume of aborted flower* (E 3) membranous, exaristate; *lower flowering glume of hermaphrodite flower* linear, tapering into a very long, rigid awn, *upper flowering glume* minute, only as long as the lodicules (E 5). *Grain* terete on back, embryo prominent on face. *Culm* simple or branched at the base.

Andropogon Ischæmum, Linn. Sp. Plant. p. 1480; Gren. et Godr. Fl. de Fr. iii. 465; Woods, Tour. Fl. p. 395.

Habitats.—(B.) From Kew Gardens, September 7, 1867. (A, C, D, E.) From Mentone, where I gathered them in the winter and spring of 1867.

Remarks.—There has been much discussion and some confusion about the structure of the hermaphrodite spikelets in Andropogon, some authors affirming and others denying the presence of a second flower or part of a second flower within each pair of outer glumes. There are distinguished authorities on either side, and among those who failed to see any trace of a second flower was M. Parlatore himself. After a careful study, I find that the rudiment of a second flower is always present (in the species figured), but that the minute upper flowering glume is often wanting in some of the spikelets, even though fully developed in others on the same spike, and this reduces the number of parts in the deficient spikelets. According to the view taken here, we must regard the sessile spikelets of Andropogon (in the European species) as composed of two flowers, one of these being reduced to a single membranous flowering glume lying within and next to the lower outer glume. In some of the extra-European species this second flower is found to present two flowering glumes and containing stamens. The species of Andropogon figured are nearly the only European representatives of the genus, and are almost peculiar to the Mediterranean and the East.

EXPLANATION OF PLATE LXXIII.—Fig. A 1, lower outer glume of the sessile spikelet. A 2, upper outer glume of the same. A 3, a flowering glume belonging to aborted lower flower. A 4, the flowering glume of the hermaphrodite flower. A 5, one of its two lodicules. A 6, one of its 3 stamens. A 7, its ovary. Fig. B 1, upper outer glume of the sessile spikelet. B 2, lower outer glume of the same. B 4, flowering glume of aborted lower flower. B 3, lower flowering glume, and B 5, upper flowering glume of hermaphrodite flower. B 6, one of the two lodicules. B 7, ovary and styles. Fig. C 1, upper outer glume of the sessile spikelet. C 2, lower outer glume of the same. C 4, flowering glume belonging to its aborted lower flower. C 3, lower flowering glume, and C 5, upper flowering glume of the hermaphrodite flower. C 6, one of its lodicules. C 7, its ovary (the stamens are omitted). C 8, a complete hermaphrodite flower, taken out of the sessile spikelet. Fig. E 1, lower outer glume of the sessile spikelet. E 2, upper outer glume of the same. E 3, the flowering glume belonging to its aborted flower. E 4, lower flowering glume of its hermaphrodite flower. E 5, upper flowering glume of the same. E 6, one of the two lodicules. E 7, stamens, and E 8, styles and ovary of the same. E 9 and E 10, two views of the grain. All the Figures magnified.

Plate 74

PLATE LXXIV.

HELIANTHEMUM TUBERARIA, Mill.

Natural Order CISTINEÆ.

GEN. CHAR.—*Sepals* 5, the two outermost smaller. *Stamens* many, all fertile. *Ovules* orthotropous. *Style* usually slender, incurved at apex, rarely wanting (as in *H. tuberaria*, Mill. and *H. guttatum*, Mill.). *Ovules* orthotropous, attached at the base. *Capsule* 1-celled, or incompletely 3-celled. *Seeds* destitute of a raphe.

SPEC. CHAR.—*Flowers* usually in a lax panicle, formed of two racemes, more rarely simple; buds nodding, flowers suberect when expanded; *bracts* and upper stem-leaves glabrous, ovato-lanceolate acuminate. *Calyx* glabrous, of 5 ovato-lanceolate, and two outer linear sepals. *Petals* longer than calyx, truncato-flabelliform, yellow and without spots at their bases. *Filaments of stamens* yellow. *Ovary* densely clothed with stellate and simple hairs. *Stigma* sessile, capitate. *Capsule* shorter than the calyx by one-half. *Leaves* ovato- or elliptico-lanceolate, attenuate below into a petiole sheathing at base, densely clothed with long and rather silky hairs, 3-nerved. *Flowering stems* ascending, developed on the lateral shoots, while the central shoot is barren, the rootstock, which is short and woody, bears the dead leaves of previous years, and is perennial.

Helianthemum Tuberaria, Mill. Dict. No. 10; Gren. et Godr. Fl. de Fr. i. 173; Woods, Tour. Fl. p. 35; Ardoino, Fl. des Alpes Maritimes, p. 47; *Tuberaria vulgaris*, Willk. Icones Plant. Hisp., &c., ii. p. 70.

HABITAT.—Behind the Hotel Bellevue, Cannes, where the specimens figured were gathered by the Rev. S. Henning, who kindly favoured me with them, April 30, 1871.

REMARKS.—This plant has been described by some botanists as the type of a distinct genus, principally on account of its sessile stigma.

Willkomm distinguishes three varieties of *H. Tuberaria*, Mill. which he names *Tuberaria vulgaris*—viz., β. *lanata*, lower leaves, even when adult, clothed with long white hairs on either side. γ. *suffruticosa*. Branches of rootstock 1-2½ inches long, erect, almost 4-sided, blackish and covered with the bases of the old leaves, forming a low shrub (*suffrutex*). Leaves of the rosettes furnished with a petiole almost equalling the limb; limb harsh, being covered above with small, sparse stellate hairs. *Helianthemum lignosum*, Sweet, t. 46 (figure taken from a remarkably luxuriant cultivated specimen). δ. *alpestris*. Rootstock perpendicular, thick, woody, branches short, forming a tuft resting on the ground. Leaves of the rosettes small, 6-8 lines long, stems short.

For the species as a whole he gives the following distribution : " This species inhabits almost the entire Mediterranean zone, except the southeastern portion, and is especially abundant in the southern parts of the Spanish peninsula, where it grows from the seashore up to 4000 feet.

The common form grows in Portugal, Spain (Galicia, Seville, Gibraltar, Valencia, and Catalonia), Provence (Cannes, Grasse, Toulon, and the Islands of Hyères, Avignon, Nîmes, and Montpellier), Corsica (near Ajaccio and Bastia), on the hills of southern Sardinia, near Nice, in the Duchy of Lucca, on Monte Pisano, in the kingdom of Naples (Otranto, Scilla, Lecce, near Naples), Sicily (near Messina and Girgenti), in Greece, and in Northern Africa (near La Calle); var. β. on hills in Algeria (near La Calle); γ. in Serra de Cintra; δ. on highest ridge of the Serra di Foia in the Portuguese province of Algarve, and in the province of Leon at Puerto de Manzenal."

Botanists should look out along the coast for *Tuberaria globulariæfolia*, Willk. (l. c. p. 71) which resembles *T. vulgaris*, but has a dark spot at base of petals and purple black filaments, and the lower leaves are spathulate ovate or ovato-lanceolate, petiolate. This is the plant figured by Curt. (Bot. Mag. t. 4873) as *H. Tuberaria*, and is the *Tuberaria perennis*, β. *globulariæfolia* of Spach; it is found in southern Portugal. Our plant gained its name of *Tuberaria* from having been found in places noted for truffles.

EXPLANATION OF PLATE LXXIV.—Fig. 1, calyx after the petals have fallen, magnified. Fig. 2, one of the inner sepals magnified, and fig. 3, the same of the natural size. Fig. 4, ovary and three stamens, the others removed, magnified. Fig. 5, ovary when more developed, magnified. Figs. 6 and 7, hairs from the same. Fig. 8, inner face of one valve of capsule of the natural size. Fig. 9, ovule and its long funicle, with the embryo laid bare, magnified.

Plate 7

PLATE LXXV.

(A) FUMANA viscida, Spach, a. thymifolium ; (B) F. viscida β. Barrelieri, Willk.; (C) F. viscida γ. juniperina, Willk.; (D) F. lævipes, Spach.

Natural Order Cistineæ.

Gen. Char.—*Sepals* 5, the two outermost smaller. *Stamens* many, filaments submoniliform, the outermost destitute of anthers. *Style* slender, enlarged above ; *stigma* (in all the species found in the Riviera district) subtrilobed, lobes fringed and papillose. *Ovules* anatropous, attached slightly below apex ; *embryo* more or less curved. *Capsule* almost 3-celled. *Seeds* furnished with a raphe.

(A.) Spec. Char.—*Peduncles* from ½ to 2 as long as calyx, the lowest distant from any bract, the others in the axils of linear bracts half as long as themselves. *Flower* very small. *Stigma* spreading into 3 large fringed lobes prolonged above into a conical tuft. *Leaves* and stipules in upper part of stem linear-lanceolate, of equal length and apparently forming a spreading whorl ; in lower part of stem and barren shoots shortly elliptic, resembling those of *Thymus vulgaris*, L., having the edges much recurved, stipules much shorter, triangular, all the leaves subsessile. *Stems* woody below, short, tortuous. Plant glutinous pubescent all over.

Fumana viscida, Spach, a. *thymifolium*; *F. viscida a. vulgare*, Gren. et Godr.? Fl. de Fr. i. 174 ; *H. thymifolium*, Pers., Ench. ii. 79 ; Woods, Tour. Fl. p. 36 (part); *H. glutinosum*, Pers. Ardoino, Fl. Alpes Mar. p. 48 (part.)

(B.) Spec. Char.—*Peduncles* nearly twice as long as calyx, glandular, pubescent, the lowest distant from any bract, the others in the axils of lanceolate bracts half as long as the peduncles. [*Flower* intermediate in size between those of the vars. *juniperina* and *thymifolium*, judging from Willkomm's figure.] *Leaves* linear, pubescent, covered with short hairs, some of which are glandular, considerably longer than the stipules, ascending or spreading, subsessile. *Stems* stiff and suberect.

Fumana viscida β. Barrelieri, Willk. Ic. Plant. Hisp. ii. p. 160, tab. CLXIV. Fig. 2.; *F. viscida a.* Gren. et Godr.? Fl. de Fr. i. p. 174; *Helianthemum Barrelieri*, Ten. Prod. Fl. Neap. p. 31 ; *H. glutinosum*, Pers. Ardoino, Fl. Alpes Mar. p. 48 (part); *H. thymifolium β. Barrelieri*, Woods, Tour. Fl. p. 37.

(C.) Spec. Char.—*Peduncles* 2-3 times calyx, springing from axila of minute triangular, rarely longer and stipulate bracts, covered with glandular pubescence, as are the adjacent parts of stem. *Calyx* covered with mixed glandular and simple hairs. *Stigma* spreading into three deeply fringed lobes, prolonged above into a central conical tuft. *Seeds* deeply pitted, pale yellow brown. *Leaves* linear, acute, distinctly petiolate, subglabrous, with a few equidistant, marginal, and one terminal bristle-shaped hair. *Stipules* erect, always shorter than the leaf.

Fumana viscida γ. juniperina, Willk. Ic. Plant. Hisp. ii. 159, tab. CLXIV. f. 3; *F. viscida*, Spach, γ. *juniperifolium*, Gren. et Godr. Fl. de Fr. l. 175; *Helianthemum juniperinum*, Woods, Tour. Fl. p. 36; *H. glutinosum*, Pers. Ardoino, Fl. Alpes Mar. p. 48 (part).

(D.) SPEC. CHAR.—*Peduncles* filiform, about three times calyx, glabrous, lowest peduncles opposite to bracts, upper axillary. *Bracts* short, about ⅓ peduncle, ciliate. *Calyx* covered with long hairs, many of which are glandular. *Corolla* smaller than that of *F. viscida* γ. *juniperina*, Willk. *Stigma* as in *F. viscida* γ. *juniperina*. *Seeds* deeply pitted, dark brown. *Leaves* and stipules linear-setaceous, glabrous, alternate, dark glaucous green, sessile. *Stem* woody below, very slender, much branched, ends of young branches nodding and pubescent.

Fumana lævipes, Spach, Ann. Sc. Nat. sér. 2, vi. 359; Gren. et Godr. Fl. de Fr. i. p. 174; Woods, Tour. Fl. p. 36; Ardoino, Fl. Alpes Mar. p. 48.

HABITATS.—(A.) Near Grimaldi, Mentone, Ap. 20, 1867; (B.) L'Hermitage, Hyères, May 6, 1868; (C.) Montegrosso, Mentone, April 7, 1871; (D.) near Grimaldi, Mentone, Ap. 17, 1871. All collected by myself.

REMARKS.—The four plants figured on this plate may all be found growing together at Mentone near Pont St. Louis, especially near the Child's Cross, and above Dr. Bennet's garden. Had space permitted I should like to have added a figure of *Fumana Spachii*, Gren. et Godr., which also abounds at Mentone, and which is easily distinguished by its exstipulate leaves. The genus *Fumana* is characteristic of the Mediterranean region, and almost limited to it in a northward direction; *F. procumbens*, Gren. et Godr., alone extending into central and northern France, Switzerland, Germany, and the Swedish island of Gothland (Nyman*). Most closely allied to *Helianthemum*, it is distinguished from it by the curious character of the barren outer filaments destitute of anthers, and by its seeds, the funicle or suspending cord of which is united to the seed coat in the greater part of its length, thus forming the projecting ridge containing the feeding vessels of the embryo, called the raphe. It is a curious, though perhaps merely coincident, fact that the stamens in wild hybrids between *Cistus salvifolius*, L., and *C. monspeliensis*, L., are frequently reduced in size or absent. All the parts of the flower are, however, liable to suffer, and I have seen an entire bush of × *Cistus monspeliensi-salvifolius* covered with minute flowers destitute of stamens, not larger than those of *Fumana viscida* γ. though either parent of this hybrid has large flowers.

All the forms represented are common between Marseilles and Genoa,

* 'Sylloge,' p. 226.

but perhaps *F. viscida* γ. is more frequently met with than the rest, and this variety displays some at least of its flowers throughout the day, while those of the others are extremely fugitive. This is especially the case with *F. lœvipes*, Spach, the petals of which have usually fallen before eleven o'clock on bright days. Differences of this kind are probably of great importance in the maintenance of races and species, as affording checks to intercrossing and favouring the visits of distinct insects.

EXPLANATION OF PLATE LXXV.—Fig. A 1, pistil of *F. viscida a. thymifolium*, magnified. B, *F. viscida* β. *Barrelieri*. C 1, and C 2, fertile and barren stamens of *F. viscida* γ. *juniperina*, magnified. C 3, pistil of the same. C 4, seed, and C 5, section of the same, magnified. C 6, pair of leaves and their stipules with young leaves in their axils, magnified. C 7, leaf of the same, magnified. D 1, pistil of *F. lœvipes*, magnified. D 2, leaf and stipules of the same, magnified.

PLATE LXXVI.

VIOLA ARBORESCENS, L.

Natural Order VIOLACEÆ.

GEN. CHAR.—See description of Plate LIV. Part III.

SPEC. CHAR.—*Flowers* round in outline; *peduncle* glabrous or puberulent; *bracts* exceedingly minute. *Sepals* triangulari-lanceolate, acuminate, membranous and ciliate at edge, appendages about ½ length of limb, subtriangular, irregularly toothed. *Petals*, the upper two very broadly obovate, on short claws; the two lateral broadly and obliquely obovate, beardless; lowest petal ovate, shorter than the lateral petals, entire or subemarginate at apex, spur short, rounded, scarcely exceeding calyx-appendages, hairy within. *Stamens*, the two lowest ones spurred, spur short, decurved, linear-compressed, bent back against one of the anther cells. *Stigma* rounded, orifice very small, scarcely projecting. *Capsule* oblongo-obtuse, glabrous. *Leaves* linear-lanceolate, sometimes having a short tooth on either side above the middle, glabrous or puberulent; *stipules* linear, puberulent. *Stems*, almost woody, forming dense tufts decumbent below. *Plant* perennial, flowering in October, destitute of stolons.

Viola arborescens, Linn. Sp. Plant. p. 1325; Gren. et Godr. Fl. de Fr. i. 182; Woods, Tour. Fl. p. 40.

HABITAT.—St. Cyr (department Var), on the neck of the promontory known as Les Baumelles, where I gathered this plant in profusion, and in excellent condition, on October 28th, 1868.

REMARKS.—In order to find the nearest relations of this curious and rare violet, we must look for *Viola decumbens*, Linn. fil. at the Cape, and *V. arborea*, Forsk. in Yemen, both of which closely resemble *V. arborescens*, L. In France, besides the St. Cyr station, *V. arborescens* grows near Toulon (Gren. et Godr.); on the Montagne de la Clappe, near Narbonne; and at Ste. Lucie. In Spain it is found near the shore along the whole eastern line of coast, attaining an altitude of 1000 to 2500 feet;[*] also in Portugal (Nyman) and in Algiers; while a marked variety having much broader and many toothed leaves (γ. *serratifolia* of De Candolle's Prodromus, a synonym of *V. suberosa*, Dsf.) grows also in Algiers, and near Mogador and Tetuan.

I am puzzled to account for the fact that M. Boissier, and all the collectors in Spain and Algiers whose specimens I have seen, give March or April as the date of flowering, and a specimen in the Kew herbarium

---

[*] Boiss. Voy. Esp. ii. p. 71.

from near Narbonne was also gathered in April, while I found this plant in full blow at St. Cyr in October, and MM. Grenier et Godron state that September is the usual flowering season for this species. It is true, however, that M. Boissier found plants near Gibraltar which in April exhibited only the fertile minute flowers, similar to those produced in autumn by our common sweet and dog violets, so that these individuals may perhaps have had full-sized flowers in the preceding autumn. I should like to know whether *V. arborescens*, L., habitually produces flowers of full size twice in the year and also the minute fertile flowers.

It would appear that the present plate is the first illustration yet published of this remarkable plant, with the exception of the quaint figure given in Jacques Barrelier's *Icones plantarum per Galliam et Italiam Obs.*, edited by Antoine de Jussieu in 1714.

EXPLANATION OF PLATE LXXVI.—Fig. 1, calyx and part of peduncle, magnified. Fig. 2, one of the two lowest (anterior) stamens, magnified. Fig. 3, ovary, style, and stigma, magnified. Fig. 4, ripe capsule, of the natural size. Fig. 5, a leaf of the natural size. Fig. 6, stipules and lower part of the leaf, magnified.

A

A. 1

A. 2

B. 1

B. 2

C. 1

C. 2

D. 1

D. 2

E. 1

F. 1

G. 1

PLATE LXXVII.

POLYGALA NICÆENSIS, Risso.

Natural Order POLYGALEÆ.

GEN. CHAR.—*Calyx* usually persistent (deciduous in *P. Chamæbuxus*, L.), of five sepals, three of which are similar, small and herbaceous, while the two lateral ones are petaloid (the wings), and much larger. *Corolla* gamopetalous, formed by the union of 1 anterior and 2 posterior petals, the anterior petal largest, concave, more or less deeply fringed. *Stamens* 8, united to the petals and forming two lateral bundles; *anthers* basifixed, unilocular, the two cells having become confluent, opening by one (or two in *P. Chamæbuxus*, L.), short cleft or pore at the apex (Caruel*). *Capsule* borne on a more or less distinct carpophore, two-celled, strongly compressed laterally. *Seeds* bearing a three-lobed aril at the hilum.

SPEC. CHAR.—*Flowers* in a lax raceme, becoming very long at maturity of fruit. *Intermediate bract* often exceeding the young buds, making the raceme appear comose, often also shorter than or only equal to them; *lateral bracts* (or *bracteoli*), symmetrical, subelliptic acuminate, or, more rarely, subovate obtuse, as long as the pedicel of the expanding bud. *Wings* (*lateral sepals*) most frequently large and broadly obovate subobtuse, sometimes obovato-elliptic, or even elliptic acute, central and 2 lateral nerves branched and anastomosing. *Capsule* oblong-obovate, or obcordate, notched or subentire at apex, narrower than or as broad as the wings, narrowed below into a very short carpophore about one-sixth its length or less. *Leaves* broadly lanceolate, the lower ones obovate, usually pubescent, edges slightly recurved. *Stems* flexuose, much branched, decumbent below, springing from a woody perennial stock.

Polygala nicæensis, Risso, Fl. de Nice, p. 54; Woods, Tour. Fl. p. 43; Ardoino, Fl. Alp. Mar. p. 54; *Polygala rosea*, Gren. et Godr. (non Dsf.), Fl. de Fr. i. 194.

HABITAT.—Garavan Valley, Mentone, where I gathered these specimens on April 7th, 1871.

REMARKS.—The common European species of *Polygala* are with few exceptions variable in their characters, and more or less doubtfully distinct. Thus, I believe that if a really large series of specimens of *Polygala nicæensis*, Risso, *P. vulgaris*, L., and *P. comosa*, Schk., were minutely examined in a fresh state, it would become quite impossible to assign to any one of these forms characters absolutely distinctive; for in a quite limited investigation which I made myself, I found that the actual length,

---

* T. Caruel. *Polygalacearum italicarum conspectus* in Nuovo Giorn. Bot. Ital. i. 19 (March, 1869).

and comose appearance of the bracts, is not unfrequently much reduced in *P. nicæensis*, when it could scarcely be said to differ from *P. vulgaris*. The supposed distinctive points of *P. nicæensis* and *P. vulgaris* may be compared as follows :—

| *P. nicæensis.*—Intermediate bract longer than young buds; lateral bracts equalling or exceeding the pedicel, subelliptic; wings larger. | *P. vulgaris.*—Intermediate bract shorter than young buds; lateral bracts only half as long as pedicel; suborbicular ; wings smaller. |

But on fixing my attention on any one of these details, and applying the test of absolute and relative measurement, it became evident that though in the majority of cases these characters are really distinctive, in a large minority there is variation which tends to bridge over the interval between the two species. For example, in ten plants of *P. nicæensis* examined, the intermediate bract varied in length from 3 to 5½ millimetres; though the length of these bracts in each individual plant was constant with but slight exceptions.

Similarly the lateral bracts varied in shape and length ; and the wings in proportion and absolute length and breadth as represented in the figures D, E, F, G, where the wings of the plant D measure 9 mm. by 7 broad ; of E, 8 by 5½ ; of F, 10½ by 6¼; of G, 9½ by 5½.

In the monograph of the Italian species of *Polygala* by Prof. Caruel, mentioned above, *P. nicæensis* is described as having pink flowers, but it is a curious fact that at Nice, Mentone, and Cannes the flowers of this plant are almost always deep blue, and very rarely white or pink, while from Oneglia to Genoa its flowers are, as Prof. Caruel describes them, of a fine deep pink. This pink-flowered form, which appears to me to constitute a distinct subvariety, characterized by its straight and more dense racemes and upright habit, is found as far westward as the banks of the Nervia river near Ventimiglia, about nine miles east of Mentone. It is, I believe, the plant which has been taken by some botanists for *Polygala anatolica*, Boiss.,* with which it has no connexion, *P. anatolica* having a carpophore but slightly shorter than the ripe capsule itself, and the tube of the corolla projecting beyond the wings, while the carpophore of *P. nicæensis* is extremely short, and the corolla tube quite included. *P. rosea*, Dsf., which has frequently been confused with *P. nicæensis*, is much more closely related to *P. anatolica*, Boiss., and *P. major*, a very handsome plant, and has the carpophore nearly half as long as the capsule, and petals which project considerably. *P. Preslii*, Spreng., a Sicilian species, has narrow ovate lanceolate wings, almost recalling those of *P. monspeliaca*, L., and a more exsert corolla than *P. nicæensis*, which it otherwise closely resembles, and of which it may indeed be a variety, being labelled by M. Boissier in his herbarium *P. nicæensis*, var. *Preslii*.

* I am greatly indebted to MM. Boissier and Reuter for a series of specimens of several species of *Polygala*, and among them *P. anatolica*, *P. rosea*, Dsf., and *P. Preslii*, Spreng.

Whether *P. nicæensis* is found further to the westward than Fréjus, I do not know; Prof. Caruel states that it is found in the neighbourhood of Nice, in Northern Tuscany, and to the east of the Apennines from Ancona to Rimini.

EXPLANATION OF PLATE LXXVII.—Fig. A 1, upper part of inflorescence in bud, magnified. A 2, an expanding bud, with bracts, magnified. B 1, intermediate bract, and B 2, one of lateral bracts of a distinct plant, magnified. C 1, C 2, the same from another plant. D 1, one of the wings from a distinct plant, and D 2, a capsule, both of the natural size. E 1, F 1, and G 1, wings from so many separate plants, of the natural size.

PLATE LXXVIII.

POTENTILLA SUBACAULIS, Linn., var. *albicans*, Moggridge.

Natural Order ROSACEÆ.

GEN. CHAR.—See description of Plate X. Part I.

SPEC. CHAR.—*Flowers* yellow, 2-3 in short panicles furnished with amplexicaul bracts. *Calyx segments* ovato-lanceolate, broader than the linear-oblong, subobtuse outer segments. *Petals* entire or sub-emarginate, obovate (obcordate and deeply emarginate in specimen in Linnæus' Herbarium), slightly longer than the sepals. *Carpels* deeply reticulate. *Leaves* trifoliolate, grey or felted on either side with mixed stellate and simple hairs; leaflets more or less broadly obovate, or in the outermost and oldest leaves cuneato-obovate, toothed in upper half only, teeth short, the terminal tooth often shorter than the rest; *petiole* short; *stipules* linear lanceolate, usually bi- or trifid, dilated and sheathing at the base.

Potentilla subacaulis, var. *albicans*, Moggridge; *P. subacaulis*, Linn. (part.) Sp. Plant. p. 715; Gren. et Godr. Fl. de Fr. i. 527; Woods, Tour. Fl. p. 118; *P. cinerea* γ. *velutina*, Lehm. *Revisio Potentillarum* in Nov. Act. Acad. Cæs. Leopold. Car. xxiii., suppl. (1856) p. 173-4.

HABITAT.—Eastern end of Mont Cheiron, north of Grasse; collected by my father, in the early part of May, 1870.

REMARKS.—*Potentilla subacaulis*, Linn., is one of those species, the nomenclature of which has a complicated history, all the more involved because it appears that the characters which are assigned to it, though distinctive in the majority of specimens, fail in certain portions of its range. *P. cinerea*, Chaix, is said to be distinguished from *P. subacaulis*, L., by its laxer habit, longer flowering stems, usually 5-foliolate and less densely felted leaves, the terminal tooth of which is smaller than the lateral ones and the narrower stipules of the leaves of the rosette. However, on examining a really large series of specimens at Geneva, in London, and at Kew, I found that no one of these characters is absolute. Lehmann, in his *Revisio Potentillarum*, quoted above, limits his species *P. subacaulis* to the form found in Siberia, but it is certain that Linnæus included under this name the south European plant and probably also the *P. cinerea* of Chaix.

There are two specimens of *P. subacaulis* in Linnæus' own Herbarium, differing in habit, and possibly from distinct stations (no habitats are given, and Linnæus has only written " 22 subacaulis" underneath) ; the one compact and with foliage resembling the plant I figure, but having a flowering stem more than twice as long as the leaves and larger emarginate petals, while the other has a much looser habit of growth and elongated decumbent flowering stems. Linnæus especially mentions the rocks of Sainte Victoire near Marseilles, as a habitat for *P. subacaulis*, and I have seen in the British Museum specimens from thence, bearing the date 1783, and named in Dr. Solander's hand, which closely resemble the Mont Cheiron plant. I also possess specimens from this neighbourhood, collected by M. Roux, and communicated by M. de Mercey, but one of these has a much laxer habit, the leaves being less densely felted and having longer petioles. Now specimens of the Marseilles plant from Sainte Victoire, Tête de Carpiagne, and Roquefavour were sent by M. Alphonse Derbès to M. F. Schultz of Wissembourg, who pronounced them to belong to his *P. Tommasiniana*, a species which he has created at the expense of *P. cinerea*, Chaix.

I have, unfortunately, not been able to realize the supposed distinctive characters of *P. Tommasiniana*, though I have examined M. Schultz' own specimens* from Monte Spaccato, near Trieste, as well as many others from the same district. I am doubtful whether the reticulation of the carpels, which was very marked in the four ripe carpels of the Cheiron plant, is or is not a constant character.

In the three cases in which I have been able to examine carpels of *P. subacaulis* from Illyria, the carpels were quite smooth; but those of plants from the Rhône valley showed every stage from deep reticulation to almost complete smoothness. The Siberian plant, judging from Lehmann's figure and specimens from the Altai, gathered by Ledebour, in the Kew Herbarium, very nearly approaches our Mont Cheiron plant in foliage and habit, but the petals appear to be emarginate.

P. subacaulis var. *albicans* is found in the Department des Alpes Maritimes on Mount Lachen above Seranon (Ardoino), on the Mont Cheiron, and over the north-eastern border of the department in the Pesio Valley, where it was gathered by Balbis, who communicated it to J. Gay.

Explanation of Plate LXXVIII.—Fig. 1, a panicle entire, of the natural size. Fig. 2, calyx, slightly enlarged. Fig. 3, a mature carpel of the natural size, and fig. 4, the same magnified. Figs. 5 and 6, leaves of the natural size; fig. 7, the apex of a leaflet magnified; and fig. 8, two leaflets of one of the older leaves, magnified. Fig. 9, one of the simple, and fig. 10, one of the stellate hairs from a leaf, magnified.

---

* No. 257 of his *Herbar. Normale*, Cent. 3, where he refers to his *Archives de Flore*, in which, at p. 273, I find the following rather vague allusion: " Ayant observé des différences rémarquables entre la plante que je donne ici, et celle que j'ai donné dans la prémière centurie, sous le nom de *P. incana* Mœnch, je l'ai décrite sous le nom de *P. Tommasiniana* dans un journal botanique allemand."

PLATE LXXIX.

PEUCEDANUM venetum, Koch.

Natural Order UMBELLIFERÆ.

GEN. CHAR.—"Perennial, rarely annual. *Leaves* pinnately, or 3-nately compound. *Umbels* compound, many-rayed; bracts few or 0; bracteoles many or 0; flowers white, yellow, or pink, often polygamous. *Calyx-teeth* 0, or small. *Petals* with an inflexed, often 2-fid point. *Disk-lobes* small; margin often expanded, undulate. *Fruit* ovoid, oblong or sub-orbicular, much dorsally compressed, commissure very broad; carpels flattish, lateral primary ridges of each forming flat contiguous wings, dorsal and intermediate filiform; vittæ 1-3 in each interstice. *Seed* nearly flat."—Hooker, *Student's Flora of Brit. Is.* p. 168.

SPEC. CHAR.—*Flowers* white, arranged in several umbels forming a paniculate inflorescence; *rays of umbel* rough along inner face. *Bracts of general involucre* 5-8, spreading or sub-reflexed. *Styles* long, at length reflexed and equalling or exceeding one-third of fruit. *Leaves* dark-green and shining above, 3-pinnate, leaflets triangular-ovate, ultimate divisions lanceolate or linear lanceolate, scabrous at edge.

Peucedanum venetum, Koch., Syn. Fl. Germ. ed. i. p. 304; Gren. et Godr. Fl. de Fr. i. 689; Woods, Tour. Fl. p. 152; Ardoino, Fl. Alp. Mar. p. 158.

HABITAT.—Mentone valley, Mentone, in a shady part, where I gathered the specimen figured on December 11th 1865. (Figs. 3, 4, and 5 are, however, from a specimen in the Kew Herbarium, gathered at Prades in the Pyrénées Orientales.)

REMARKS.—*Peucedanum venetum*, Koch., is nearly related on the one hand to *P. alsaticum*, Linn., which is distinguished, however, by its yellowish flowers, short styles, and smooth rays of umbel, and on the other to *P. austriacum*, Koch., in which the bracts of the general involucre are much more numerous and reflexed.

There are specimens of *P. venetum* in the herbarium at Kew, gathered near Prades in the Pyrénées Orientales (name of collector and date not given), this forming an isolated and hitherto unrecorded station on the extreme west of its range; from Outre Rhône, about half-way between St. Maurice and Martigny, where it was detected by E. Thomas, who communicated the specimen to M. J. Gay; and from the Val Vestino in the Western Tyrol (Porta).

MM. Grenier and Godron give but one station for *P. venetum* in France—viz. the Chartreuse de Valbonne. It appears that this species is found here and there along the southern slopes of the Swiss and Italian Alps, reaching as far eastward as Croatia. In the department of the Alpes Maritimes it is local, having been found near Nice, Sospello, La Giandola, and lately near Bezaudun in the arrondissement of Grasse by the Abbé Consolat. Professor De Notaris states that it is frequently found in woods and meadows between Genoa and Nice.

EXPLANATION OF PLATE LXXIX.—Fig. 1, a flower magnified. Fig. 2, an immature fruit magnified. Fig. 3, outer face of one of the mericarps of a ripe fruit magnified. Fig. 4, inner or commissural face of the same magnified. Fig. 5, transverse section of the same magnified. Figs. 3, 4, and 5 are from a dried specimen in the Kew Herbarium, gathered at Prades.

1.

2.

3.

Plate LXXX.

CAMPANULA isophylla, Morett.

Natural Order Campanulaceæ.

Gen. Char.—"Perennial, rarely annual herbs. *Radical leaves* usually petioled, cauline alternate. *Flowers* spiked or racemed, white, blue, or lilac. *Calyx-tube* ovoid or subglobose; limb flat or folded at the sinus. *Corolla* campanulate or rotate, 5-lobed. *Stamens* 5, epigynous, filaments short, bases broad, dilated; anthers linear, free. *Ovary-cells* 3-5, opposite the sepals; style clavate, with rows of deciduous hairs opposite the anther-cells, stigmas 3-5 filiform. *Capsule* ovoid or turbinate, 3-5 celled, cells dehiscing below or above the calyx limb by pores or valves. *Seeds* usually flattened."—Hooker, *Student's Flora of Brit. Is.* p. 225.

Spec. Char.—*Flowers* large, lilac, or rarely white, in compact terminal corymbs. *Calyx lobes* lanceolate acute ("sometimes toothed," DC.*), glabrous or finely pubescent, about half as long as corolla, spreading in flower, erect in fruit. *Corolla* rotato-campanulate, lobes from half to two-thirds as long as tube. *Style* exsert. *Capsule* subturbinate, dehiscing by pores at the base, erect. *Seeds* minute, elliptic, compressed, shining yellow brown. *Leaves* roundish cordate, crenate, more or less deeply toothed, the lowest often reniform, all being otherwise nearly uniform in shape, petiolate, pubescent. *Stems* suberect.

Campanula isophylla, Morett. Append. ad Schow. prosp. fl. Ital. p. 22; *C. floribunda*, Viv. Fl. Lyb. append de Fl. Ital. p. 67; Woods, Tour. Fl. p. 239.

Habitat.—Promontory of Caprazoppa between Noli and Finale, where I gathered the specimens figured Oct. 26, 1870.

Remarks.—This beautiful *Campanula* is only known to grow along the small strip of coast, about two-and-a-half miles in length, from the promontory of Caprazoppa to near the little town of Noli, and is a singular parallel to the case of *Convolvulus subatius*, Viv. (see Plate LXI. Part III.), which is also peculiar to this district. It is very difficult to understand why it is that *Campanula isophylla*, Morett, should be so restricted in its area, and the more so as it produces large quantities of seed, which, as I have proved, germinate freely, and which are so minute that they might be transported to any distance by the wind alone, or adhere, without causing inconvenience, to the feet of small birds when wetted by hopping in dewy grass, &c.

Campanula fragilis, a plant from Naples and Sicily, frequently cultivated

* DC., Prodr. vii. 476.

in gardens near London, is the nearest ally of the present plant, but is distinguished from it by having diffuse ascending stems, and smaller ovate or lanceolate stem leaves.

It is a curious fact in the distribution of plants, that all the species belonging to the section of perennial Campanulas which have erect capsules and usually rotate corollas, sixteen in number,[*] are almost limited to Italy, Greece, and the immediately adjacent regions, except that *C. argentea*, Lam., extends as far as Armenia, and *C. macrorhiza*, Gay, into Spain.

EXPLANATION OF PLATE LXXX.—Fig. 1, dehiscent capsule, showing the erect calyx-lobes, magnified. Figs. 2 and 3, leaves from other specimens than those figured, of the natural size, 2 being one of the lower, and 3 the lowest leaf.

* DC., Prodr., l.c.

B

A.6.

A.1.

A

A.3.

A.4.

A.2.

B.2.

A.5.

B.1.

PLATE LXXXI.

(A) CONVOLVULUS sepium, Linn.; (B) C. sylvaticus, Waldst.
et Kit.

Natural Order Convolvulaceæ.

Gen. Char.—See description of Plate XXX. Part II.

(A.) Spec. Char.—*Flowers* large, white or pale pink; *bracts* ovate or subtriangular, cordate at base, applied against the calyx and nearly parallel to, but not overlapping, one another, becoming separated as the capsule ripens. *Peduncle* 4-angled. *Sepals* nearly equal, ovate-lanceolate. *Seeds* smooth, obtusely 2-angled. *Leaves* deeply cordate, angular or rounded at base. *Stems* long, twining. *Rootstock* whitish, fleshy, brittle, creeping underground. *Convolvulus sepium*, Linn. Sp. Plant. p. 218; Gren. et Godr. Fl. de Fr. ii. 500; Woods, Tour. Fl. p. 250; Ardoino, Fl. Alp. Mar. p. 261

(B.) Spec. Char.—Differs from the preceding in its inflated, overlapping, cordato-orbicular bracts which enclose the capsule. The sepals also are often narrower and more acuminate, and the leaves have the basal lobes usually approximate and rounded at the base. *Convolvulus sylvaticus*, Waldst. et Kit. Ic. Plant. Hung. iii. p. 290, tab. 261; *C. sylvestris*, Willd. En. pl. h. Berol. p. 202; Woods, Tour. Fl. p. 249; *C. lucanus*, Ten. Fl. Nap. Prodr. App. v. p. 9, tab. 213.

Habitats.—(A) Richmond (Surrey, England), Sept. 27, 1867. (B) Turin Valley, Mentone, Jan. 3, 1867. The specimens collected by myself.

Remarks.—I have not had the opportunity of examining any large series of fresh specimens of *Convolvulus sylvaticus*, as its time of flowering is almost over before the winter season sets in; but I suspect that any one who had the means at his disposal would find that the characters assigned to this plant are inconstant, and that it should be placed as a variety of *C. sepium*. This latter plant is well represented in the herbarium at Kew, but I could not satisfy myself that the varieties link *sepium* on to *sylvaticus*. One plant from Lexington, Kentucky, collected by Dr. Short, and named *C. sepium*, appeared to me to approach more nearly to *C. sylvaticus*: but I rather distrust this conclusion, drawn from dried specimens, and the more so as Professor Asa Gray* states that *C. sepium* is common and variable in the States, but makes no allusion to *C. syl-

* 'Man. Bot. of Northern U.S.,' p. 376.

vaticus. Choisy, in De Candolle's Prodromus,[*] however, states that *C. sylvaticus* is found in North America. This species is commonly culti- vated near London.

The range of *C. sepium* may indeed be said to be world-wide, and this plant is cited by M. Alphonse De Candolle as an example of the rule that those species which inhabit road-sides and waste places are found dis- persed over large geographical areas. M. Alphonse De Candolle gives[†] the following distribution for *C. sepium:* Europe, as far as Sweden; in Kasan; the Caucasus, Southern Siberia; Australia, New Zealand, Chonos Archipelago, Chili, California, Oregon, Newfoundland, Azores, Algiers. *C. sylvaticus* grows at Constantinople (Choisy), in the Banat, Italy and Sicily, Dalmatia, Macedon and Thrace, and it is common along the Riviera from Genoa to Mentone, mixed with *sepium*, and I have seen a specimen in M. Thuret's herbarium, which he collected near Golfe Jouan, between Antibes and Cannes.

The specimens in Linnæus, herbarium, which are remarkably fine and well-preserved, under which Linnæus has written "Convolvulus sepium, 2, Alger," appear to me to be, without doubt, characteristic examples of *C. sylvaticus.* There is no question, however, but that Linnæus described the form common in Northern Europe; and, indeed, he gives no extra- European habitats for *C. sepium* even in the second edition of the Species Plantarum.

EXPLANATION OF PLATE LXXXI.—Fig. A 1, the bracts and peduncle of the natural size. A 2, a bract of the natural size. A 3, a stamen, magnified. A 4, the ripe capsule, enclosed in calyx and bracts, of the natural size. A 5, the same, deprived of calyx and bracts. A 6, a seed of the natural size. Fig. B 1, bract of the natural size. B 2, a division of the calyx, of the natural size.

[*] Vol. ix. p. 433.   [†] Geog. Bot. i. p. 573.

PLATE LXXXII.

(A) CONVOLVULUS UNDULATUS, Cav.; (B) C. PENTAPETALOIDES, Linn.

————————

Natural Order CONVOLVULACEÆ.

GEN. CHAR.—See description of Plate XXX. Part II.

(A.) SPEC. CHAR.—*Flowers* axillary, subsessile, the very short peduncle furnished with linear bracts at its base. *Calyx* of 5 ovate-acuminate, ciliate, herbaceous, unequal sepals, longer than the peduncle. *Corolla* funnel-shaped, shortly 5-lobed, nearly white with a ring of pale pink above the yellow throat. *Ovary* hidden in long silky hairs. *Root leaves* on petioles as long or longer than the ovato-elliptic blade; the leaves next above are spathulate, and the distinction between limb and petiole gradually dies out, so that the upper leaves become sessile, subauriculate, broadly ovate. *Plant* annual, leaves and stems covered with long fine hairs, not twining.

Convolvulus undulatus, Cav. Ic. iii. 30, tab. 277, f. 1; *C. evolvuloides*, Dsf., Fl. Atlant. tab. 49; Wood's Tour. Fl. p. 250.

(B.) SPEC. CHAR.—*Flowers* axillary, on long peduncles, recurved in fruit, furnished midway with linear bracteoles. *Calyx* of 5 ovate, subequal, subglabrous sepals, scarious in upper half of length, scorched at tip. *Corolla* spreading, deeply 5-lobed, finely pubescent on back, lilac, with yellow throat, paler outside. *Ovary* glabrous. *Root-leaves* on petioles equalling or exceeding the elliptic limb; the leaves next above are sub-spathulate, and the distinction between petiole and limb gradually dies out, so that the upper leaves become sessile, auriculate, oblongo-lanceolate. *Plant* annual, leaves finely pubescent, stems covered with short adpressed pubescence, not twining.

Convolvulus pentapetaloides, Linn. Syst. Nat. ed. 12, iii. p. 229; Wood's Tour. Fl. p. 250.

HABITATS.—(A) San Remo, April 19, 1871. (B) San Remo, April 19, 1871. Both gathered by the Rev. Boscawen Somerset, and M. Panizzi.

REMARKS.—*Convolvulus undulatus*, Cav., is a rare species, and, as far as I can learn, San Remo is the only station for this plant in Italy, the other recorded habitats being Sicily (Nyman), central and southern Spain, Barbary, Egypt (Willkomm and Lange*), and in Tunis at Sbiba, west of Kairwan (Desfontaines); there are also specimens from near Saïda in Algiers (Cosson) in the Kew Herbarium. The plant figured by Sibthorp and Smith in the *Flora Græca* (tab. 198), gathered on the island of Cyprus corresponds with *C. undulatus* except in colour, the corolla being there represented as of an uniform lilac in its upper third. *C. pentape-*

———————

* Prodr. Fl. Hisp. ii. 518.

PLATE LXXXIII.

OMPHALODES verna, Mœnch.

Natural Order BORAGINEÆ.

GEN. CHAR.—"*Calyx* 5-partite, or sub-5-partite. *Corolla* broadly funnel-shaped or almost rotate, the tube being extremely short, tube rarely elongate cylindrical, throat closed with obtuse scales. *Stamens* 5, included in tube, anthers elliptic, longer than filament. *Style* simple, emarginate at the capitate apex. *Ovary* 4-partite. *Achenes* 4, depressed, affixed by base or side to the style, having a membranous, cup-shaped, broad margin inflexed above. *Seeds* . . . . obliquely ascending, horizontal or pendent; cotyledons flat, fleshy, obovate, much shorter than the radicle." DC. Prodr. x. 158. [Translated: the italics are mine.]

SPEC. CHAR.—*Flowers* in twin, terminal, few-flowered-racemes, having 1-3 leaves at their bases but no bracts; *fruiting pedicels* recurved. *Calyx* covered with greyish, adpressed pubescence, lobes lanceolate acute. *Corolla* lilac in bud, afterwards deep blue, paler outside, rather more than half as long again as calyx. *Carpels* obovate, having the raised rim entire, pubescent when young. *Leaves* ovate or the lower ones cordato-ovate, finely pubescent, whitish below when young, few on stem. *Rootstock* creeping, throwing out rooting stolons sometimes a foot long.

Omphalodes verna, Mœnch. Meth. p. 420; Gren. et Godr. Fl. de Fr. ii. 538; Wood's Tour. Fl. p. 252.

HABITAT.—San Romolo near San Remo, collected by my father on May 3rd, 1868, and March 24th, 1871.

REMARKS.—This charming plant should be seen fresh gathered from, or growing in, its wild haunts to be fully appreciated. In our old-fashioned English gardens, where it goes by the name of Venus' Navel-wort, it has, not unfrequently, spread as a weed in shady and overgrown places, where it becomes drawn and weakly, but near San Remo it rivals in beauty the Alpine Forget-me-not (*Myosotis alpestris*, Schmidt), itself.

In France there are but two recorded habitats for *Omphalodes verna*, Mœnch, namely at Russy-Montigny near Villars-Cotterets in North-eastern France (Aisne), and in the neighbourhood of Lyons. Between Marseilles and Genoa I know of no other habitat for this plant except the one near San Remo, and among the Apennines and higher hills near Genoa, where it is said* by Prof. De Notaris to be frequent. It is stated by Nyman† to grow also in Lombardy and central Italy, Austria, Carinthia, Styria, and the Tyrol, Holland and Belgium (but as a doubtful native), Hun-

* Dutrs. Rep. Fl. Lig. p. 319.　　　　† Sylloge, p. 87.

gary, Croatia, Transylvania, and the Crimea. Professor Nyman and other authors have stated that *Omphalodes verna* is found in many parts of Spain, but MM. Willkomm and Lange, in their lately published volume of their *Prodromus Floræ Hispanicæ*, distinctly say that this species does not grow wild in Spain.

EXPLANATION OF PLATE LXXXIII. Fig. 1, the entire corolla, and fig. 2, the same cut vertically in half, magnified. Fig. 3, one mature and three imperfectly developed achenes with the style, magnified. Fig. 4, a leaf of the natural size.

NOTE.—Fig. 3, is copied from that given on p. 178 of MM. Le Maout and Decaisne's Traité de Botanique.

PLATE LXXXIV.

BALLOTA SPINOSA, Lk.

Natural Order LABIATÆ.

GEN. CHAR.—" *Calyx* almost funnel shaped, tube 10-nerved, teeth 5-10, dilated at the base and either forming an orbicular spreading limb, or an oblique limb in which the anterior portion is connate and elongate. *Corolla tube* sub-included, having a transverse hairy ring ; limb bilabiate, upper lip erect, oblong, sub-concave, emarginate at apex. *Stamens* ascending under the hood. *Anthers* exsert from the tube of the corolla, approaching one another in pairs, 2-celled, cells finally divaricate, subdistinct. *Style* bifid at apex, lobes subulate. *Achenes* obtuse and not truncate at apex." Bentham in DC. Prodr. xii., 516-7. (Translated : the italics are mine.)

SPEC. CHAR.—*Flowers* 1-3, in axillary cymes, *bracts* spine-shaped, straight. *Calyx* pubescent, funnel-shaped with spreading limb formed from the dilated bases of the 5-10 spinescent teeth. *Corolla* white, decidedly longer than calyx, the upper lip densely covered with long hairs. *Leaves* pubescent, green on either face, ovate, petiolate, entire or incisodentate.

Ballota spinosa, Lk., Handbk. p. 457 ; Gren. et Godr. Fl. de Fr. ii. p. 695 ; Ardoino, Fl. Alp. Mar. p. 302 ; *B. frutescens*, Wood's Tour. Fl. p. 295.

HABITAT.—Near Gourg de l'Ora, Mentone, where it was collected by my father, Nov. 7, 1865.

REMARKS.—*Ballota spinosa*, Lk., is only known to grow in the department of the Basses Alpes at Entrevaux, and in that of the Alpes Maritimes, at the following points ; Saorgio, Breglio, Sospello, between Castellar and Castiglione, between Ste. Agnès and Gorbio ; at Eze, Nice, Levens, Villars, Le Bar, and St. Arnoux ; it attains its easternmost limit at Ventimiglia (Ardoino). The common Black Horehound (*B. nigra*, L.) is a familiar member of this genus, but belongs to a distinct section. The section in which *B. spinosa* is placed, is characterized by having the long spiniform bracts which give the present plant so strange an appearance. Mr. Bentham* enumerates five species of *Ballota* which have " subulate, stiff, spiniform bracts and the limb of the calyx 5-10 toothed," but of these only one other, *B. integrifolia*, Bth., growing in the island of Cyprus, is European ; the others come from India, Persia, and Arabia Felix.

It is interesting to remark that *B. integrifolia*, which, though distant, is geographically the nearest closely-related neighbour of our own

B. spinosa, is also structurally more nearly connected with it than with any other species, and differs from it only in its "longer branches, usually quite entire leaves, stouter and generally recurved spines, and longer calyces, the limb of which is larger and constantly 5-toothed," (Bentham). The flowering season for *B. spinosa* is in the summer, and it is only rarely that some few plants may be found in bloom in October and November.

EXPLANATION OF PLATE LXXXIV.—Fig. 1, a flower with bracts, part of the stem and the bracts of the opposite cyme in which the flowers are undeveloped ; the leaves are removed : of the natural size. Fig. 2, stamens, style, and corolla deprived of its upper lip, magnified. Fig. 3, a stamen, magnified. Fig. 4, an achene, magnified.

PLATE LXXXV.

ANDROSACE Chaixii, Gren. et Godr.

Natural Order PRIMULACEÆ.

GEN. CHAR.—*Calyx* 5-toothed or deeply cleft. *Corolla* cup- or funnel-shaped, the tube shorter than the calyx and contracted at the throat, which is almost closed by a ring of minute scales. *Style* short. *Capsule* 5-valved splitting from base to apex. *Seeds* usually few, about 5 in each capsule, sometimes many.

SPEC. CHAR. *Peduncles* glabrous above, slightly pubescent at base. *Bracts* subglabrous with ciliated edges, lanceolato-obovate or obovate, sometimes with 2 lateral teeth, each bract having a small pouch-like appendage at the base of the midrib. *Calyx* glabrous except for a few ciliæ, lobes ovate acute, shorter than the 5-sided tube, considerably enlarged in fruit and then broader than long. *Corolla* pink, 1-2 as long as calyx, lobes obtuse, entire. *Capsule* half as long again as calyx, valves narrow lanceolate. *Seeds* large, few, elliptic, convex on one face and nearly flat on the other, shagreened. *Leaves* oblongo-lanceolate, shortly toothed above, minutely pubescent or subglabrous, forming a single rosette. (I have seen a single, abnormally developed specimen in which there were small imperfect lateral rosettes.) *Plant* annual or biennial.

Androsace Chaixii, Gren. et Godr. Fl. de Fr. ii. 458; Ardoino Fl. Alp. Mar. p. 310; *A. septentrionalis*, Vill. (non L.) Hist. Plant. Dauph. ii. 281.

HABITAT.—Near Briançonnet, at 2800 ft. elevation, in the extreme north-west of the department of the Alpes Maritimes; the flowers collected by my father, May 5, 1870, and the ripe capsules by Mr. J. Orr, during the following summer.

REMARKS.—The genus *Androsace* can scarcely be limited by description so as to distinguish it from *Primula*, and Professor H. G. Reichenbach says[*] that *Androsace* differs from *Primula* only in its "usually short tube of the corolla with throat crowned with minute prominences, and its usually few-seeded capsule, splitting from base to apex." The curious, minute, pouch-like enlargements at the bases of the bracts in *Androsace Chaixii* are said by Reichenbach to be found also in *A. lactea*, L. This singular and exceptional character reappears in *Primula*, thus affording additional evidence of the close affinity of these two genera, the involucral bracts in *P. farinosa*, L., being slightly saccate at the base, while those of *P. Munroi*, Lind.,[†] a plant from Northern India, are prolonged into distinct and prominent spurs.

[*] l.c. Fl. Germ. xvii. p. 16. [†] Bot. Reg. xxii. tab. 15.

Androsace Chaixii, Gren. et Godr., is stated by the authors of the species to grow near Gap, at La Baume, near Sisteron, on Mont Ventoux, and at Castellane in the old department of the Var.

In the Alpes Maritimes it grows near Briançonnet, on Mont Cheiron, and Mont Lachen, above Séranon, and at Le Brec d' Utelle (Ardoino). The general range of this species has not as yet been ascertained, but MM. Grenier and Godron thought that they recognised in a plant, sent by M. Fischer from southern Siberia, the true *A. Chaixii!* The annual, or at most biennial, growth of this plant readily distinguishes it from those species which, like *A. lactea*, L., for example, form branched and matted tufts of rosettes. *A. septentrionalis*, L., and *A. elongata*, L., approach *A. Chaixii* very closely, but appear to be separated from it by the character of the calyx, which is not enlarged round the fruit, and the more pubescent habit of their peduncles and pedicels.

EXPLANATION OF PLATE LXXXV.—Fig. 1, calyx, pedicel, and bract, magnified Fig. 2, calyx enclosing the dehiscent capsule, of the natural size, and fig. 3, the same, magnified. Fig. 4, dehiscent capsule, magnified. Fig. 5, a seed of the natural size, and figs. 6 and 7, the same magnified. Figs. 8 and 9, involucral bracts magnified. Figs. 10 and 11, leaves of the natural size.

Plate 56

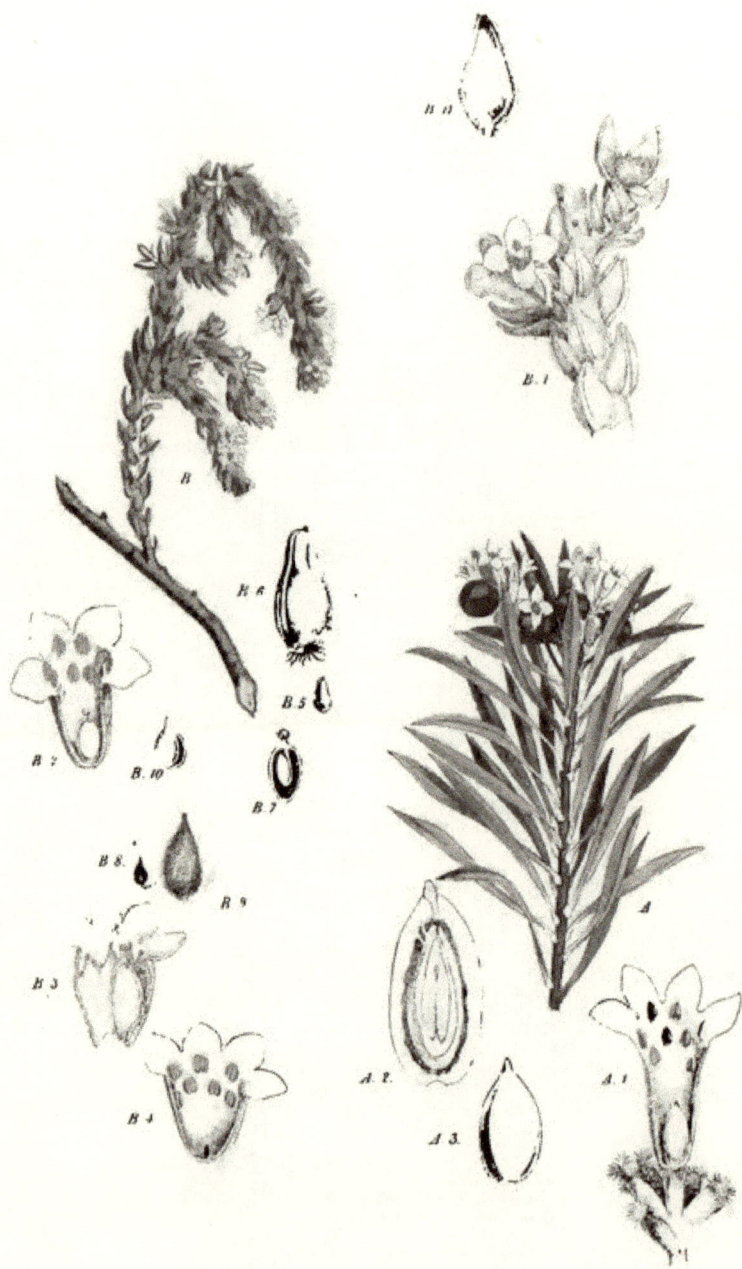

PLATE LXXXVI.

(A.) DAPHNE Gnidium, Linn.; (B.) THYMELÆA HIRSUTA, Endl.

Natural Order THYMELACEÆ.

(A.) GEN. CHAR.—*Flowers* hermaphrodite, tetramerous. *Perianth* coloured (more rarely green), tubular or funnel-shaped, deciduous, or more rarely persistent, tube continuous, limb regular, 4-partite, spreading, destitute of scales at the throat. *Hypogynous disk* obsolete or minute, annulate, sometimes forming a very short cup or dimidiate. *Anthers* 8, inserted in a double row in the throat, subsessile, oblong, subincluded. *Style* terminal, very short or none, stigma capitate. *Berry* fleshy or coriaceous, naked, or almost dry, and enclosed for some time in the calyx. *Seed* nut-like, testa crustaceous. *Albumen* none or little. *Embryo* fleshy, cotyledons plano-convex. Meisner in DC. Prodr. XIV. p. 531.

(A.) SPEC. CHAR.—*Flowers* in ebracteate racemes forming a terminal panicle. *Pedicels* covered with a dense white tomentum, the uppermost hairs of which form a sort of fringe. *Perianth* silky outside, lobes shorter than the tube. *Berry* fleshy, red. *Leaves* subcoriaceous, linear lanceolate acuminate, densely clothing the branches. *Stem* upright, woody, branched, perennial. *Daphne Gnidium*, Linn. Sp. Plant. 511; Gren. et Godr. Fl. de Fr. iii. 60; Woods, Tour. Fl. p. 522; Ardoino, Fl. Alp. Mar. p. 327.

(B.) GEN. CHAR.—*Perianth* urceolate or funnel-shaped, quite persistent or at last deciduous. *Hypogynous scales* 0. *Style* lateral or rarely terminal. *Fruit* a dry, nut-like drupe. [The other characters are as in *Daphne*.]

(B.) SPEC. CHAR.—*Flowers* 2-5 on capitate enlargements of the summit of short lateral branches, sessile, destitute of bracts. *Perianth* pale yellow, pubescent outside, about as long as the leaves, lobes slightly shorter than the tube. *Leaves* thick, evergreen, convex and smooth on back, concave and covered with white-felted hairs on face, imbricate. *Stems* much branched, forming a low bush, the younger branches of which are pendulous and covered with felted white hairs. *Thymelæa hirsuta*, Endl. Gen. Suppl. 4, ii. p. 65; Ardoino, Fl. Alp. Mar. p. 328; *Passerina hirsuta*, Linn. Sp. Plant, ed. i. p. 559; Gren. et Godr. Fl. de Fr. iii. 63; Woods, Tour. Fl. p. 331.

HABITATS.—(A.) gathered by me at Mentone, Nov. 27th, 1866; (B.) collected by Mr. Stainton on the Isle Ste. Marguerite, off Cannes, on March 12th, 1867, who kindly brought me a supply of specimens.

REMARKS.—The genera *Daphne* and *Thymelæa* seem to be scarcely separable, the terminal style and usually succulent berry of the former being the only distinctive features which are tolerably absolute. The genus *Passerina* is limited by Prof. Meisner to four species, all from the Cape. The extreme range of *Daphne Gnidium*, L., extends on the south and west to Algiers, Teneriffe and Palma in the Canaries, on the north to the shores of the western Riviera and Dalmatia, and on the east to Greece. It is commonly found along the shore between Marseilles and Genoa, but exact data are wanting. The distribution given by Meisner (DC. Prod.) for *Thymelæa hirsuta*, Endl., is as follows :—around all the Mediterranean coasts, in Spain, Portugal, Sardinia, Corsica, the Balearic Islands, southern France, at Nice, on the Ligurian shore, in Istria, the kingdom of Naples, Sicily, Zante, at Constantinople, in Crete, Syria, Egypt, Arabia petræa, and Algeria. Four varieties are also described. α. *vulgaris* (the form here figured). β. *polygalæfolia*, Endl., with ovate leaves tomentose on either side, at last glabrescent on back ; from near Marseilles. γ. *rotundifolia* with suborbicular ovate, very obtuse leaves, as broad as long, glabrous on back ; from near Collioure. δ. *angustifolia*, leaves narrow, ovato-oblong or lanceolate, somewhat acute canaliculato-concave above, with apex often inflexed, glabrous on back ; found near Carthage, Marseilles, Naples, in Sardinia, Sicily, Algeria, Crete, and Egypt. *Thymelæa hirsuta* has apparently almost attained a diœcious condition, some plants* bearing male flowers, as at B 4, with only occasional hermaphrodite flowers, while others produce, almost exclusively, female flowers, as at B 3, and others again the hermaphrodite flowers of B 2. I was able last April (thanks to the kindness of the Rev. S. Henning) to examine specimens from nine distinct bushes. The results may be briefly given as follows :—

Bush No. I., 12 flowers examined all ♀; No. II., many flowers all ☿; No. III., 15 flowers ♂ and 2 ☿; No. IV., many flowers all ☿; No. V., 4 flowers ☿ and 2 ♂; No. VI., 14 ♀; No. VII., many flowers all ♀; No. VIII., 7 flowers ♂; No. IX., 6 flowers ☿. Thus there were three bushes of which all the flowers examined were female, one in which all were male, three all hermaphrodite, and two in which there were both hermaphrodite and male flowers. It would be curious to learn, from the examination of a really large series of specimens, whether male, female, and hermaphrodite flowers are ever found together on the same bush.

Between Marseilles and Genoa *Thymelæa hirsuta* grows near Marseilles, Hyères, St. Raphael, on the islands of Ste. Marguerite and St. Honorat, on the promontory of Antibes, and between Spotorno and Vado, west of Savona (De Notaris).†

EXPLANATION OF PLATE LXXXVI.—Fig. A 1, a flower and pedicels from which the flowers have fallen, magnified. A 2, a longitudinal section of the berry, magnified. A 3, the embryo, magnified. Fig. B 1,

* The figures B 1, and B 2, by an oversight bear the letter B, as if the flowers came from the plant B. In reality, they came from distinct plants, and should have been numbered C 1, and D 1, respectively.

† "Rep. Fl. Lig." p. 389.

part of a branch with leaves and flowers, magnified. B 2, an hermaphrodite flower split open, magnified. B 3, a female, and B 4, a male flower, split open, from distinct bushes, magnified. B 5, the drupe, of the natural size, and B 6, the same, magnified. B 7, the ovary divided longitudinally, magnified. B 8 and B 9, a seed* of the natural size, and the same magnified. B 10, the embryo. B 11, upper face of a leaf, magnified.

---

* In figs. B 8, and B 9, the seed is erroneously represented as being furnished with a long free funicle, attached at the base : it is, in fact, suspended from near the apex by a short funicle, communicating with the base of the seed by a raphe, which forms a very obscure ridge along one side.

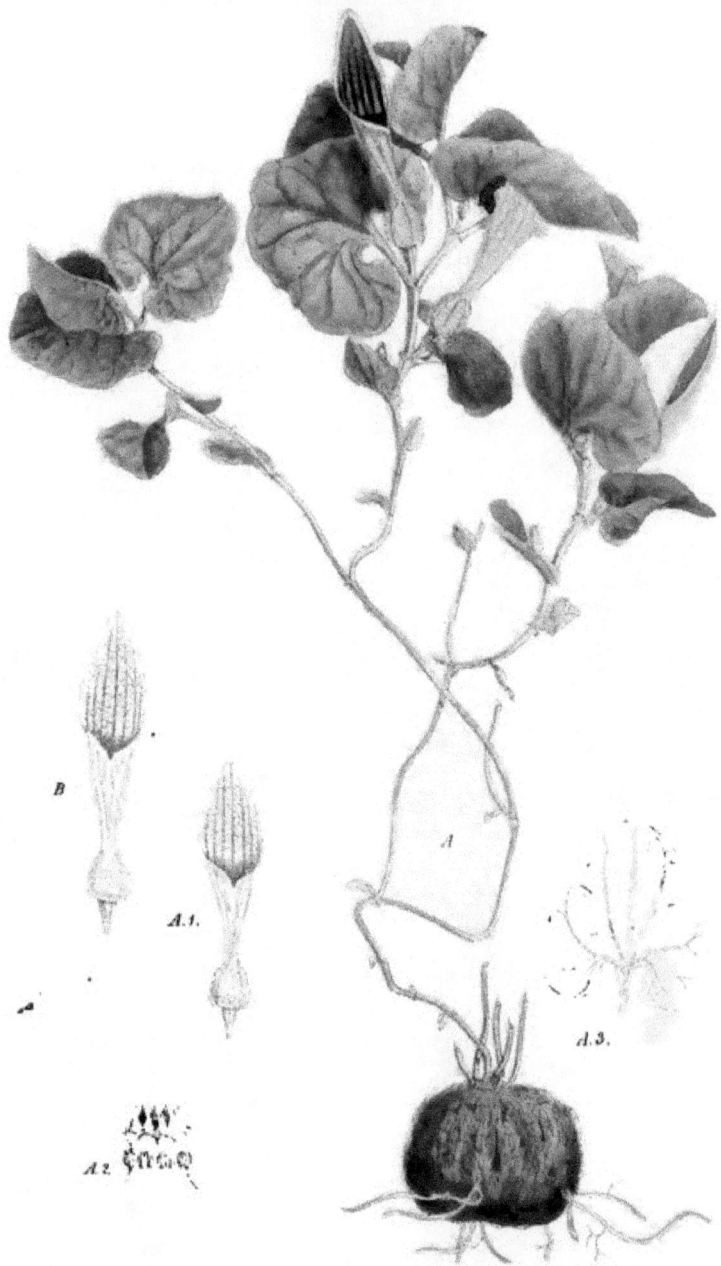

B

A.1.

A

A.2

A.3.

PLATE LXXXVII.

ARISTOLOCHIA PALLIDA, Willd.

Natural Order ARISTOLOCHIACEÆ.

GEN. CHAR.—See description of Plate LXIV. Part III.

SPEC. CHAR.—*Flowers* variable in colour, pale yellowish-green, often with a dark purple-brown blotch in the throat, the lip more or less covered with purplish reticulate markings inside; *peduncles* very short, about one-sixth of an inch long. *Perianth tube* clavate, slightly longer or shorter than the ovate or oblong-ovate, acute or obtuse lip. *Stigmatic lobes* at first erect, conical, distinct, afterwards confluent (when a kind of deli-quescence sets in). *Leaves* glabrous, broadly cordate, slightly emarginate, basal lobes usually approximate; *petiole* varying from $\frac{1}{4}$ to $\frac{3}{4}$ inch or more long. *Rootstock* subglobose.

Aristolochia pallida, Willd. Sp. Plant. iv. 162; Woods, Tour. Fl. p. 324; Ardoino, Fl. Alp. Mar. p. 330.

HABITAT.—Gathered by my father on the eastern end of Mont Cheiron (arrondissement de Grasse, Alpes Maritimes), on May 6th, 1870.

REMARKS.—*Aristolochia pallida*, Willd., appears to have no character which absolutely distinguishes it from *longa*, Auct.* except the roundish shape of its rootstock, as opposed to the fusiform or subcylindrical root-stock of the latter. The following are the distinctive peculiarities of leaves, flowers, and stems assigned by M. Duchartre,† who says that *A. pallida* is distinguished from, " though most closely allied to, *A. longa*, by its paler colour, more delicate habit, subglobose tuber, usually unbranched stem, leaves shorter in proportion to breadth, having longer petioles, the colour of the flower, the shorter and subclavate tube, and finally by the longer obtuse lip," but these characters break down one by one when applied to a large series of specimens.

However, the rootstocks appear to be markedly distinct, and it would appear that in *A. longa* at least this feature is accurately reproduced by seed, for I have received, thanks to the kindness of Dr. D. Moore of Glas-nevin, five three-year-old seedling plants of *A. longa*, in all of which the rootstocks were about as long as my hand, fusiform, and as thick as my little finger, tapering below to the size of whipcord.

A. pallida is well distinguished from *A. rotunda*, L., by its distinctly stalked leaves, and if treated as a variety it should be placed, as it is pro-visionally in M. Boissier's herbarium,‡ under *A. longa*.

---

* See description of Pl. LXIV. Part III.   † In DC. Prodr. xv. 487.
‡ The label is thus given, " *A. longa*, L., var. *radice globosa*."

A. pallida has lately been discovered by my father on the Cime d'Ours, a mountain close to Mentone, and has also been found in the department of the Alpes Maritimes along the chain of Mont Cheiron, by M. Marcilly, at about 4000 feet on the north slope in cultivated ground, near the Bastide de Gerbières at Roquesteron, and the Bastide du Poux, also on the southern slope where the road crosses some rough ground between Haut Thorenc and Mas, and again on the same slope above Coursegoules. M. Marcilly has moreover discovered this species not far from Agay, west of Cannes, and M. Ardoino states that it has been gathered near Toretta Revest, also called Tourrette, in the valley of the Vesubie, N.E. of Mont Chauve ; and last spring a specimen was gathered with its rootstock near Sospello, at the entrance of the valley leading to Molinet, by M. E. Burnat.

I possess a specimen collected by Dr. Rostan near Marseilles (Pâturages sous Marseille, No. 29 of his *Exsiccata Pedemontana*), the north-westernmost point of its range, and MM. Huet and Shuttleworth have discovered this species in the neighbourhood of Toulon on the Ste. Baume and at Collobrières, and M. Huet has again found it in the wooded district known as Les Maures, near Le Luc, to the east of Toulon, between Hyères and Vidauban. Eastward it recurs in mountain pastures above Genoa,[*] in several parts of Italy, in the Canton Tessin, the Tyrol and Carnia, Dalmatia, Greece, and the island of Zante, Croatia, Hungary, and Transylvania (Nyman).[†]

I have also seen specimens in M. Boissier's herbarium, from Alma Dagh, or Mount Amanus on the northern frontier of Aleppo, gathered by Balansa, and from near Madrid, gathered by M. Reuter.

EXPLANATION OF PLATE LXXXVII.—Fig. A 1, perianth of plant figured, of the natural size. A 2, column and stamens of the same, magnified. A 3, leaf of same, natural size. B, flower from a distinct plant of the same species, of the natural size.

[*] Dntrs. Rep. Fl. Lig. p. 392.   [†] Syll. Fl. Eur. p. 330.

NOTE.—The plants figured fairly represent the average form of the perianth in the numerous specimens which I have received from the arrondissement of Grasse, and from the Cime d'Ours, near Mentone, but the lip is sometimes much shorter in proportion to the tube, being for example, but ½ of the tube in a specimen from near Sospello, and but ⅓ in one from Southern Italy.

PLATE LXXXVIII.

NECTAROSCORDUM siculum, Lindl.

———— ————

Natural Order LILIACEÆ.

GEN. CHAR.—*Perianth* almost bell-shaped, of 6-8* dissimilar segments which are united at the base, and form a discoid tube in which the ovary is partly immersed ; *outer segments* oblongo-elliptic, having a nectariferous depression at the base; *inner segments* broadly ovate, abruptly contracted below into a narrow claw. *Stamens* 6-8,* inserted on the tube, nearly similar in length and form ; *filaments* simple, subulate ; *anthers* sub-oblong, bilobed at base, fixed near the base at the back, cells parallel, introrse. *Ovary* depressed, sub-disciform, about one-third immersed in the perianth tube (having nectariferous apertures at the apices of the thickened sutures, Parl.), 3- or, imperfectly, 8-celled.* *Style* subulate, gynobasic. *Capsule* 3-4 valved,* enclosed within and adherent at the base to the persistent segments of the perianth, having a small circular pit on the back. *Seeds* compressed, angular, testa black, minutely punctate.

SPEC. CHAR.—*Pedicels* enlarged below the flower, recurved during the expansion of the flowers, erect in fruit. *Scape* 2-3ft. high, surrounded at the base by a dilated sheathing leaf, which completely encloses it when young, and divides along the central line of its surface in the upper third of its length to admit of the passage of the umbel of buds then enclosed in their spathe. *Leaves* broadly linear, channelled, and strongly keeled, so as to be almost triquetrous, spirally twisted above, making 2-3 turns, all, with the exception of the sheathing leaf described above, free from the base of the scape. *Bulb* simple.

Nectaroscordum siculum, Lindl. Bot. Reg. (1836), p. 1913, tab. 1912 ; Parl. Fl. Ital. ii. 584 ; Ardoino Fl. Alp. Mar. p. 380 ; *Allium siculum*, Uc. Pl. ad Linn. opus addend. No. 7. (Parl.) ; Gren. et Godr. Fl. de Fr. iii. 212 ; Woods, Tour. Fl. p. 371.

HABITAT.—Petite Suivière du Malpey à l'Esterel, on the southern slope of a mountain forming part of the Esterel range, named in Cassini's map the Montaigne de Montuby, at 1200 ft. above the sea. Collected by my father, in flower on May 9, 1869, and in seed on Nov. 9, of the same year.

REMARKS.—This plant seems quite sufficiently distinct from any known species of *Allium* to allow of its being placed in a distinct genus, but no one who has gathered or dissected fresh specimens will fail to be reminded

* The increased number of parts is only found in the 2 or 3 first expanded and central flowers of well developed inflorescences, and not at all in small and few-flowered umbels.

of the relationship, by its strong garlic odour, and its power of making the eyes water. *Nectaroscordum siculum*, Lindl., has only been discovered as yet in this one station among the Esterels, where the dense brushwood and the distance from any roads make the search for specimens one of great difficulty, and also among the mountains of Sicily, and on Monte Gennargentu in Sardinia. I have, however, seen a plant in M. Alphonse de Candolle's herbarium bearing the following label :—"No. 2870, Allium, in Mte. Nâlkou (Assyrie), M. Aucher Eloy, 1836," which is attributed to this species, and which, as far as one may judge from specimens in an imperfect condition, appears to be indeed *Nectaroscordum siculum*. If this plant from Kourdistan[*] be really our species, we may perhaps have here an indication that this plant is a relic of the times when the east was more completely one in its vegetation with the western portion of the Mediterranean than it now is, and when the Cedars of Atlas, now distant 1400 miles from their relations, spread in more or less continuous forests to join the Cedars of Lebanon and the Taurus range, while these again were united with the Deodar Cedars of Affghanistan and the Himalaya, bridging over another space, now blank, of 1400 miles![†] Besides the partially inferior ovary, the dissimilar divisions of the perianth and the sheathing leaf or bract from which the flower scape of this plant emerges, another interesting feature in its structure, and one which appears to have hitherto escaped observation, is the difference which exists between the central and outer flowers of well-furnished umbels, in respect of the number of their parts. I find that in all stout and well grown plants of *Nectaroscordum*, the two or three central flowers have 8 divisions of the perianth, 8 stamens, and an 8-celled ovary (becoming 4-celled in the ripe capsule by abortion), while the outer flowers have their parts in multiples of 3, there being 6 divisions of the perianth, 6 stamens, and a 3-celled ovary. This difference between the innermost and the outer flowers might be supposed to be due to an excess of nourishment which is supplied to those flowers, which most nearly spring from the summit of the central axis of the plant; but this alone will not serve to explain the assumption of a fixed number of additional parts. For, if this were the case, one might expect to find the numbers of the parts proportioned to the amount of nutriment received, so that in large vigorous umbels the maximum would be reached by the central flowers, while the adjacent flowers would have their parts in sevens, eights, or sixes; but this apparently never takes place, the flowers being either trimerous or tetramerous in their plan. Among Monocotyledons the trimerous arrangement is almost universal, but among the few exceptions we find *Smilacina bifolia*, Roem. et Schult., and *Paris quadrifolia*, L., which have the parts of their flower either on a dimerous or trimerous plan. I do not know of any Monocotyledonous plant which has throughout its floral organs the

[*] Mte. Nâlkou is near and north of Kermandshah on the western borders of Persia.

[†] For an account of these Cedars, their affinities and distribution, see Dr. Hooker in *Natural History Review*, II. p. 11. (1862).

pentamerous arrangement so common in Dicotyledons, and it would appear that the bias of Monocotyledons is primarily towards having flowers, the parts of which are in multiples of 3, and secondarily in multiples of two. In Dicotyledons we have the central metamorphosed flower of the common carrot, *Adoxa Moschatellina*, L., in which the lateral flowers are pentamerous and the terminal flower tetramerous, and *Chenopodium rubrum*, L., in which the terminal flowers are pentamerous and the lateral trimerous with 1 or 2 stamens (Koch.).

EXPLANATION OF PLATE LXXXVIII.—Fig. 1, an inner, and fig. 2, an outer segment of the perianth of the natural size. Fig. 3, a stamen of the natural size, magnified at fig. 4. Fig. 5, the ovary, the perianth segments, and stamens being removed, with part of pedicel of the natural size. Fig. 6, transverse section of the ovary of a tetramerous flower, magnified. Fig. 7, dehiscent capsule of a tetramerous flower, slightly enlarged, and fig. 8, the same of a trimerous flower. Fig. 9, one valve of a capsule, slightly enlarged.

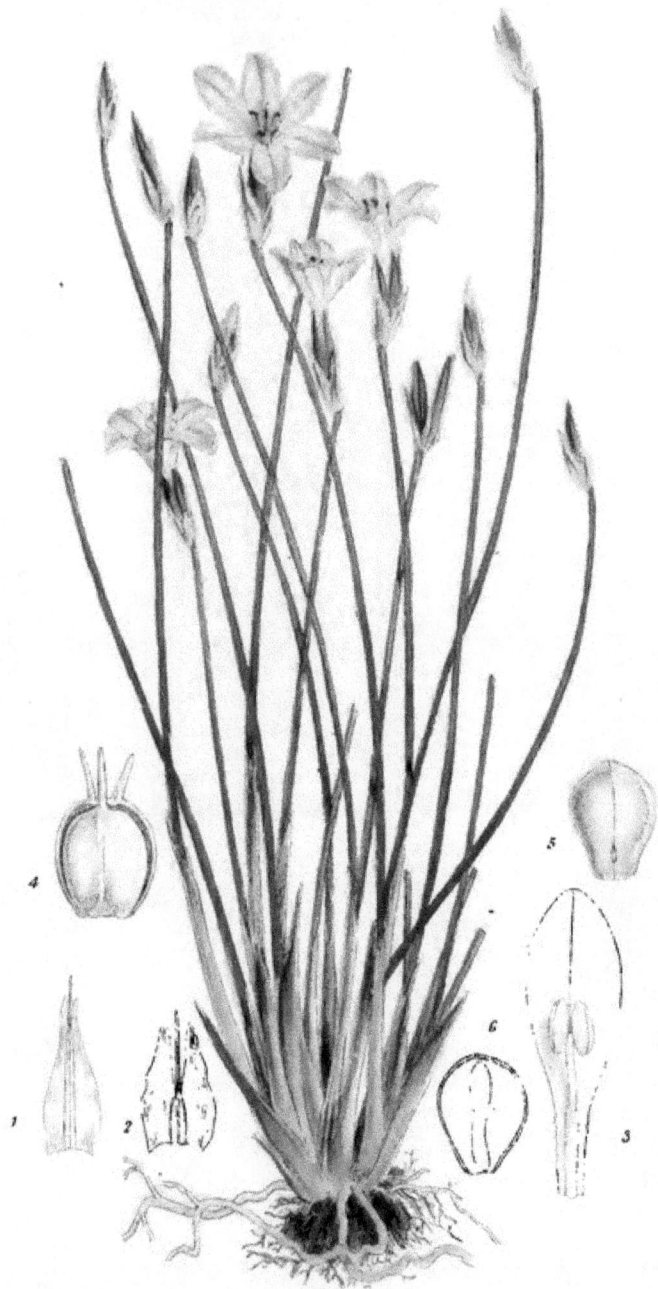

PLATE LXXXIX.

APHYLLANTHES MONSPELIENSIS, L.

Natural Order LILIACEÆ.

GEN. CHAR.—*Flowers* terminal, solitary or 2-3, having a large 5-fid bract and 3 simple bracts forming an involucre. *Perianth* petaloid, deciduous, of six divisions, distinct and spreading above, the long claws united below and forming a tube. *Stamens* 6, inserted at the top of the claws, unequal, the 3 outer ones shorter; *filaments* filiform, glabrous; *anthers* bilocular, introrse, emarginate at apex, bifid at base, fixed above the base at the back, cells parallel, united towards the middle by the narrow connective. . . . *Ovary* free, stipitate, oblongo-fusiform, trigonous, trilocular. *Ovules* solitary in the cells, inserted on the central axis by means of short funicles, amphitropal. *Style* terminal, filiform, triquetrous above; *stigma* 3-fid, enlarged into 3 recurved lobes diverging above. *Capsule* covered by the persistent bracts, membranous, trilocular, loculicidal, 3-valved, valves bearing the septa in the middle. *Seeds* solitary in the cells; *testa* crustaceous, black; embryo in the axis of fleshy albumen.—Parl. Fl. Ital. ii. 366.

SPEC. CHAR.—*The three outer bracts* are of thinner and more membranous texture than the innermost 5-fid bract, they are also distinctly nerved; all are more or less deeply bifid, the intermediate bract being most deeply divided and having a long awn proceeding from the notch; the innermost bract is less than half the size of the outermost and largest bract. *Innermost 5-fid bract* at first sight resembles a calyx, but what appears to be a tube is in reality open along one side, so that the bract can be unrolled; it is of an almost horny texture and destitute of nerves. *Flower* shortly pedicellate within the bract, divisions rather pale blue inclining to lilac, darker along midrib, obovate on long claw. *Leaves* reduced to sheaths at the base of each flowering stem. *Stems* unbranched, rush-like, tufted, springing from a short, scaly, branched, and matted rhizome. *Roots* at first white, long, similar to those of *Asparagus acutifolius*, L.

Aphyllanthes monspeliensis, L. Sp. Plant. p. 422; Gren. et Godr. Fl. de Fr. iii. 225; Woods, Tour. Fl. p. 374; Ardoino, Fl. Alp. Mar. p. 383.

HABITAT.—Mentone, from hill side above the head of St. Jacques Valley, where I gathered the specimen figured, on 12th of April, 1865.

REMARKS.—*Aphyllanthes monspeliensis*, L., the sole representative of its genus, is one of those plants whose presence in Europe it is difficult to explain. Intermediate in structure and aspect between the lily and the rush families, it appears to belong properly neither to one nor the other, and has been referred by Parlatore* to a separate family (*Aphyllanthaceæ*), where it is associated with *Borya, Alania, Johnsonia,* and *Laxmannia,* all of which are Australian, *Laxmannia* alone being also found in Timor (Meisner).

Among European plants our own Lancashire Asphodel (*Narthecium ossifragum,* Huds.) though widely dissimilar, comes nearer to *Aphyllanthes* than any other liliaceous plant; but it seems that we must seek for its true relations in Australia.

Aphyllanthes monspeliensis is found in the Mediterranean region of Europe and northern Africa. It grows in Portugal, in the southern part of Spain and France, where it reaches as far northward as Lyons and Grenoble, and to near Chambery in Savoy. It was found in Algeria by Desfontaines, and grows in Egypt also, for it is to this species that we must refer the *Dianthus prolifer* of Forskal, Fl. Ægypt. Arab. p. xxv. (Parl. l. c. p. 369). This plant abounds on banks among myrtle and heath bushes between Marseilles and Toulon, and is not uncommon in the Department des Alpes Maritimes, and thence eastwards to near Genoa. I have no record of its existence between the Esterel mountains and Toulon, but it may probably be found there also. The flowers of *Aphyllanthes monspeliensis* are very sweet during the day time.

EXPLANATION OF PLATE LXXXIX.—Fig. 1, outermost bract, and fig. 2, bract next above. Fig. 3, one division of perianth with stamen attached. Fig. 4, dehiscent capsule. Fig. 5, entire seed, and fig. 6, the same cut longitudinally in half, and showing the embryo. All the figs. represent the parts magnified; figs. 4, 5, and 6 are copied from the *Traité de Botanique* of MM. Le Maout and Decaisne, p. 584.

* l. c. p. 364.

NOTE.—In the present plate I have inadvertently represented a plant in which the divisions of perianth are broader than usual.

A.1

A.3. A.2.

B

B.1.

B.4. B.5.

B.6.

A.6.

A.5.

A.

B.3. B.2. A.4.

PLATE XC.

(A) NARCISSUS CHRYSANTHUS, DC.; (B) N. BERTOLONII, Parl.

——————

Natural Order AMARYLLIDACEÆ.

GEN. CHAR.—See description of Plate XXII. Part I.

(A.) SPEC. CHAR.—*Flowers* large and loose, yellow with orange crown, 9-12. *Scape* nearly cylindrical, not 2-edged. *Divisions of perianth* slightly longer than tube, the outer oblongo-elliptic, mucronate, the inner subobtuse. *Crown* small, cupshaped, sides upright, about one-fourth length of divisions of perianth, orifice nearly circular, entire. *Leaves* tapering from base to summit, rather deep green, broad, angularly channelled. *Narcissus chrysanthus*, DC. Fl. Fr. v. 323 ; Gren. et Godr. Fl. de Fr. iii. 258 (part); Woods, Tour. Fl. p. 360 (part); Ardoino, Fl. Alp. Mar. p. 370.

(B.) SPEC. CHAR.—*Flowers* small, yellow, with yellow-orange crown, about 5. *Scape* markedly 2-edged, not prominently ribbed, 11 in. long. *Divisions of perianth* about one-third shorter than tube, the outer oblongo-obovate mucronate, the inner ovate acute. *Crown* small, cupshaped, sides upright, about one-third divisions of perianth, orifice nearly circular, entire. *Leaves* glaucous green, peculiarly narrow, evenly curved and not angularly channelled on surface 10-10½ in. long.

Narcissus Bertolonii, Parl. Fl. Ital. iii. 132.

HABITATS.—(A.) From a fresh specimen, and a photograph sent to me by Dr. Bornet, from M. Thuret's garden at Antibes, where plants of *N. chrysanthus*, DC., originally obtained from Grasse, are cultivated. Feb. 9, 1870. (B.) From plants originally obtained in terraces at San Remo, cultivated in my garden at Mentone, Dec. 13, 1869.

REMARKS.—The two species represented in this plate are allied to *N. aureus*, Lois. (Part I. Plate XXII.), but are much rarer forms. The very broad divisions and deeper colour of the flower distinguish *N. aureus* from these at a glance ; while the proportions of the divisions no less readily distinguish *N. chrysanthus* from *N. Bertolonii*. *N. chrysanthus* is only known as yet as growing in the neighbourhood of Grasse, and at Le Bar, a village to the north-east of Grasse, in the department of the Alpes Maritimes. *N. Bertolonii*, Parl. is said by Parlatore to grow in the neighbourhood of Lucca and Pisa only, but has since been discovered in tolerable abundance near San Remo. I have received from San Remo a curious hybrid, which was evidently the offspring of *N. remopolensis*, Panizzi (a subspecies split from *N. Tazzetta*), and *N. Bertolonii*, among tufts of which it was found growing by the Rev. B. Somerset. The flowers had yellowish-white divisions, prolonged as in *N. remopolensis*, with a deeper tinge of yellow at the base. The peculiar appearance of this plant is not easily conveyed by description, but when seen it plainly bore the stamp of an abnormal origin.

N. Bertolonii habitually flowers very early, coming into blossom even in November, in favourable seasons, and it can have only been some stray remainder flowers which served to fertilize or bear seed by *N. remopolensis*, a plant which does not flower habitually before February. In this case we can understand how it is that, because of their distinct seasons of blooming, hybridization should rarely take place; but in many other instances it is a profound mystery why two plants which grow together, are visited by the same insects, and occasionally do produce hybrid offspring, are not constantly intercrossed. There must be both checks which prevent, and tendencies which favour hybridization, which exist unknown to us, and this I take to afford evidence of the manifold nature of variation in plants, which is so often present though unperceived by us. For it seems probable that certain individuals of a given species have a far greater aptitude for receiving the fertilizing influence of the pollen of a distinct species than others; just as there is no doubt that some individuals are much more fertile than others of the same species when treated in the ordinary way. Careful observations repeated on the same individuals during several consecutive seasons, with a view to proving the average fecundity of each, might probably give valuable results, tending to clear up this question.

EXPLANATION OF PLATE XC.—Fig. A 1, part of a flower cut longitudinally, of the natural size. A 2, an outer, and A 3, an inner division of perianth of the natural size. A 4, a transverse section of a leaf at the base, and A 5, at the middle, of the natural size. A 6, transverse section of stem of the natural size. B 1, part of a flower cut longitudinally, of the natural size. B 2, and B 3, outer and inner divisions of perianth of the natural size. B 4, stigma and part of style, magnified. B 5, transverse section of leaf at the middle, of the natural size. B 6, transverse section of scape, of the natural size.

PLATE XCI.

(A, B, C.) ROMULEA COLUMNÆ, Seb. et M. ; (D, E, F.) R. COLUMNÆ
a DISCRETA, Moggridge.

Natural Order IRIDACEÆ.

GEN. CHAR.—See description of Plate XCIII. below.

(A, B, C.) SPEC. CHAR.—*Scape* 2-3 flowered. *Upper valve of spathe*
usually hyalino-membranous, more rarely having a narrow central line of
herbaceous tissue. *Perianth* usually very small, nearly white, with one
purplish central stripe; *segments* oblongo-lanceolate, very pale yellow-
green outside with a purplish central stripe. *Style* slightly shorter than
the stamens. *Leaves* 2-3 inches in length, stout, recurved. *Romulea
Columnæ*, Seb. et M. Fl. Romanæ Prodr. p. 18 ; Woods, Tour. Fl. p. 357 ;
Ardoino, Fl. Alp. Mar. p. 363 ; *Trichonema Columnæ*, Rchb. Fl. Excurs.
i. 83 ; Gren. et Godr. Fl. de Fr. iii. 238.

(D, E, F.) SPEC. CHAR.—*Scape* 1-flowered. *Upper valve of spathe* completely membranous. *Perianth* from half as large again to twice as large
as that of *R. Columnæ*, white or bluish-white with grey purple tinge,
and three purplish stripes; *segments* oblongo-lanceolate, the outer ones
being yellow-green on the back with three broad purplish stripes. *Style*
slightly shorter than the stamens. *Leaves* lineari-filiform, less curved
and more slender than in the species. *Romulea Columnæ a discreta*,
Moggridge.

HABITATS.—(A, B, C.) Cabrolles Valley, Mentone, where I collected the
flowering specimens on March 3rd, and the fruiting April 17th, 1870.
(D, E, F.) Cap Croisette, Cannes, collected by my father, March 23rd, 1870.

REMARKS.—I have as yet seen but half a dozen fresh specimens (all
from the Croisette at Cannes) of *R. Columnæ* var. *discreta* represented in
this plate. It has a distinct look, but I doubt whether it will prove to be
permanently separable as a species.

I have not been able myself to botanize the neighbourhood of Cannes at
the time when the species of *Romulea* are in flower, but I have reason to
think that several interesting forms might be discovered there. A single
bulb of the variety *discreta* cultivated at Mentone reproduced this spring
the characters of the plant represented at figure D, and yielded the capsule drawn in the left-hand corner of this plate. I hope to sow the seeds
next autumn, with a view to observing whether the peculiarities of the
variety are maintained. Judging from dried specimens there are slight
differences between the form assumed by *R. Columnæ*, Seb. et M. at Mentone, and the British plant from Jersey and Dawlish, these latter
having larger and pale violet and green-tinted flowers, the segments
of which overlap one another.

R. Columnæ is found in the department of the Bouches du Rhone, on the shores of the "Etang de Berre,"[*] near Toulon, Hyères, and Aigues Mortes (Gren. et Godr.), Fréjus, Cannes, Mentone, and near Pegli (De Notaris).[†] Its general range reaches its extreme northern limit at Dawlish, and follows the north-western and western coast of France, and the southern and north-eastern coasts of Spain,[‡] to reappear again at Montpellier and traverse the Riviera as recorded above, and after showing itself at several points in western and south-western Italy, Sicily, Ischia, and Corsica (Parlatore), finds its eastern limit in Greece.

EXPLANATION OF PLATE XCI.—Fig. A 1, an outer segment of the perianth of the natural size, and fig. A 2, the same magnified. A 3, an inner segment of the perianth of the natural size, and fig. A 4, the same magnified. A 5, a stamen and portion of the perianth, and fig. A 6, the same magnified. A 7, stamens and style with base of perianth, magnified. A 8, upper valve of the spathe, magnified. Figs. B and C represent other specimens of *R. Columnæ*. E 1, an outer, and E 2, an inner segment of the perianth, of the natural size. E 3, a stamen with a portion of the perianth attached, of the natural size. E 4, the style of the natural size. E 5, stamens and style with base of perianth, magnified. E 6, the upper valve of the spathe, magnified. Figs. D and F represent other specimens of *R. Columnæ* var. *discreta*.

---

* Derbès et Castagne, Cat. Pl. Bouches du Rhone, p. 153.   † Rep. Fl. Lig. p. 430.
‡ Wilk. et Lange, Fl. Hisp. Prod. i. 145.

PLATE XCII.

(A) ROMULEA RAMIFLORA, Ten.; (B) R. RAMIFLORA α CONTORTA,

Moggridge.

———————

Natural Order IRIDACEÆ.

GEN. CHAR.—See description of Plate XCIII. below.

(A.) SPEC. CHAR.—*Scape* elongate, 2-3 flowered. *Upper valve of spathe* herbaceous, with very narrow membranous margins. *Perianth* rather small, yellow in throat; *outer segments* oblong-lanceolate, yellowish outside and deep or pale lilac in upper part; inside *inner segments* lanceolate, purple in upper part, inside and out [I have seen specimens gathered at Fréjus in which the flowers were white with yellow throat]. *Style* slightly shorter than the stamens. *Leaves* 6-12 inches in length, stout, straight or recurved (the two outer leaves are represented as being bent backwards in order to bring them within the limits of the plate). *Bulb* simple. *Romulea ramiflora*, Ten., Append. ad indic. sem. H. R. N. (1827); Woods, Tour. Fl. p. 357; Ardoino, Fl. Alp. Mar. p. 363.

(B.) SPEC. CHAR.—*Scape* long, 3-4 flowered. *Peduncles* curiously contorted. *Upper valve of spathe* almost entirely membranous. *Perianth* rather small, yellow in throat; *segments* lanceolate acute. *Style* slightly longer than the stamens. *Leaves* shorter and more recurved than in A. *Bulb* enclosing one or more leaf-bearing bulbils in the outer coats. *Romulea ramiflora α contorta*, Moggridge.

HABITATS.—(A.) Carnolés, Mentone, where I gathered the specimens figured, March 16th, 1870. (B.) Cap Croisette, Cannes; collected and communicated to me by Mr. Strickland, March 23rd, 1870.

REMARKS.—The variety of *Romulea ramiflora*, Ten., which I have named *contorta*, appears to be of extreme rarity, having as yet been found only in one spot, in small quantities, on the Cap Croisette. I have received and collected large numbers of specimens of *R. ramiflora* from Mentone, Cannes, Fréjus, and Hyères, but have never before observed any marked deviation from the typical characters. Four bulbs of the variety *contorta* were placed in my garden and flowered last spring: one of these produced flowers twice as large as those represented in the figure, and the upper valve of the spathe was herbaceous with a narrow scarious edge suddenly enlarged below; two of the others resembled that just described, but had an even, wide, scarious margin, and the herbaceous nerves were separated by lines of membranous tissue. The fourth plant closely reproduced the characters of the individual represented in the plate (B). *R. ramiflora*, Ten., is found in southern Spain, Sicily, central and southern Italy, at Mentone, Cannes (on the Croisette and at St. Cassien), Fréjus, (where a white-flowered variety has been discovered by Mr.

Strickland), Hyères, and Montpellier.* It has probably been overlooked in many places, for though when in flower it is with few exceptions an easily recognised species, the season of blooming quickly passes, and besides, the perianths are rarely to be found expanded except when the sun shines directly upon them.

EXPLANATION OF PLATE XCII.—Fig. A 1, an outer, and A 2, an inner segment of the perianth, of the natural size. A 3, a stamen with a portion of the perianth attached, of the natural size. A 4, stamens and style with the base of the perianth, magnified. A 5, the style of the natural size, and A 6, the same magnified. A 7, upper valve of the spathe, magnified. Figs. B 1 and B 2, outer and inner segments of the perianth, magnified. B 3 and B 4, a stamen, of the natural size and magnified. B 5, stamens and style with base of perianth, magnified. B 6, style of the natural size. B 7, upper valve of the spathe, magnified.

* MM. Richter and Loret in Bull. Soc. Bot. de France, xiii. 245 (1866).

Plate XCIII.

(A.) ROMULEA bulbocodium, Seb. et M. ; (B, C.) romulea
ligustica, Parl.

Natural Order Iridaceæ.

Gen. Char.—*Perianth* petaloid, funnel shaped, having a short tube
and a 6-partite limb of nearly equal segments. *Spathe* of 2, opposite,
conduplicate valves. *Stamens* 3, inserted on the perianth tube; *anthers*
basifixed, extrorse. *Ovary* inferior. *Style* filiform; *stigmas* bipartite,
either branch being filiform, papillose on the upper side, and recurved.
Leaves linear, compressed, sulcate. *Scape* usually bearing two or more
peduncles.

(A.) Spec. Char.—*Scape* 1-3 flowered. *Upper valve of spathe* herbaceous,
with broad, transparent, scarious margins, acute. *Perianth* having a
short tube, and a yellow pubescent throat; *segments* lanceolate acute.
Style longer than the stamens, and but slightly shorter than the perianth-
segments. *Leaves* long, and but slightly curved; (they are represented as
being bent backwards in the figure, in order to bring them into the limits
of the plate.)

Romulea bulbocodium, Seb. et Maur., Fl. Romanæ Prodr. p. 17; Parl.
Fl. Ital. iii. 243; Woods (partly), Tour. Fl. p. 357; *Trichonema bulbocodium*, Ker. in Annals of Bot. i. 222; Gren. et Godr. Fl. de Fr. iii. 238.

(B, C.) Spec. Char.—*Scape* 3-6 flowered; *peduncles* slender. *Upper
valve of spathe* entirely scarious and transparent, obtuse. *Perianth* having
a short tube, and a white pubescent throat; *segments* oblongo-lanceolate,
or. lanceolate. *Style* exceeding, equalling, or shorter than the stamens,
which are themselves short in proportion to the perianth. *Leaves* of
medium length, suberect or recurved.

Romulea ligustica, Parl. Fl. Ital. iii. 249.

Habitats.—(A.) From Pegli, near Genoa, collected by Mrs. Tebbs,
January 31, 1867. (B, C.) Valle di Molinazze, Sestri, near Genoa, collected
by my father on April 5, 1870.

Remarks.—The form of *Romulea bulbocodium*, Seb. et M., represented
here, and that of the Genoa district generally, is a remarkably handsome
and large flowered one, and corresponds with the variety *grandiflora* of
which M. Boissier kindly sent me specimens from Syria, collected by
Gaillardot; the flowers are fully one-third larger than average specimens
of the same species from the south-west coast of France (Bordeaux, &c.)
and from Istria. I am indebted to M. Motelay for a good supply of
fresh specimens of this plant from Bordeaux. I gather that this rather
smaller and shorter leaved form is the *R. syrtica* of MM. Jordan and
Fourreau,* though it does not present the purplish colour in the upper

* Icones. Tab. cviii. p. 165.

B. B.1 B.2 B.3 A. A.1 A.2

PLATE XCIV.

(A) SERAPIAS NEGLECTA, Dntrs. ; (B) S. LONGIPETALA, Poll.

Natural Order ORCHIDACEÆ.

GEN. CHAR.—See description of Plate XVI. Part I.

(A.) SPEC. CHAR.—*Petals* subulate above, enlarged below into a broadly ovate base, concave, wavy at edge. *Lip* flesh-coloured, hairy, large, the lateral lobes projecting beyond the sepals, terminal lobe cordate acute ; guiding-plates (calli) separate and nearly parallel. *Pollen-masses* greenish-yellow. *Leaves* and *stem* free from spots. *Tubers* 2, one more or less distinctly stalked.

Serapias neglecta, Dntrs., Repert. Fl. Ligust. p. 423 ; Parl. Fl. Ital. iii. 430 ; Ardoino, Fl. Alp. Mar. p. 358.

(B.) SPEC. CHAR.—*Bracts* acuminate, exceeding flowers. *Petals* subulate above, enlarged below into an oblongo-ovate base, concave, wavy at edge. *Lip* reddish-brown, hairy ; lateral lobes short ; terminal lobe lanceolate or ovate acuminate elongate, bent backwards and adpressed against the basal half of the lip ; guiding-plates separate, nearly parallel. *Pollen-masses* dusky green. *Leaves* and *stem* free from spots. *Tubers* 2, both sessile or nearly so.

Serapias longipetala, Poll. Fl. Veron. iii. 30 ; Gren. et Godr. Fl. de Fr. iii. 278 ; Parl. Fl. Ital. iii. 424 ; Ardoino, Fl. Alpes Mar. p. 358 ; *S. pseudo-cordigera*, Moric. Fl. Venet. p. 374 ; Koch. Syn. Fl. Germ. Ed. 3, p. 661 ; *S. lingua*, Woods, Tour. Fl. p. 355.

HABITATS.—(A) Gathered by me near the Croix de Garde at Cannes, May 3, 1869. (B) gathered by me in the Mentone Valley, Mentone, April 29, 1870.

REMARKS.—The specimen of *S. neglecta*, Dntrs., figured, though small, is of about the average size for the plant as found about Cannes, but at Genoa it grows with much greater freedom, and is usually twice as tall. I am indebted to M. Huet of Toulon for specimens of this species from Lavadun and Les Maures, both in the department of the Var, these being, as far as I can learn, the westernmost points in the range of *S. neglecta*, the other known stations for which are Cannes, Mentone, where I once found a solitary specimen—Sestri, Genoa, Sarzana, Vallechio, in the Apuan Alps, Monte San Quirico, Monte Pisano, Valle d'Asciano, Melfi in the Basilicata, and Ajaccio in Corsica (Parlatore.) The apparatus for fertilization in this plant is quite similar to that in *S. longipetala*, Poll., and to that described and illustrated above (Plate XVI. Part I.) in *S. cordigera*, all three species being absolutely sterile when unvisited by insects.

I regard *S. longipetala*, Poll., as a somewhat doubtful species, as I have gathered forms at Hyères which appear intermediate in character between the plant figured and *S. cordigera*. At Cannes, Mentone, and Genoa, however, I have seen no such links, and reserve my opinion as to whether the Hyères plants may be intermediate varieties or hybrids.

Prof. Parlatore gives the following sketch of the range of *S. longipetala*, Poll.† : "It is found in France, Italy, Sicily, Corsica. . . . . Greece, on Mount Hymettus (Fraas), in Turkey, near Constantinople, in the Crimea, and grows also in Soria, near Antioch (Kotschy), and in the Caucasus (Grieseb. pl. exsicc. !)"

This plant, however, though apparently widely spread along the northern shores of the Mediterranean, is scarce at Cannes and Mentone, and though more abundant at Genoa and Hyères, is still of less frequent occurrence than some of the other species.

The five species of *Serapias* described and figured in this work, are the only ones as yet found along the western Riviera, if we exclude *Serapias triloba*, Viv., which appears to be a hybrid between *Orchis papilionacea*, L., and *Serapias cordigera*, *S. neglecta*, or *S. lingua*.‡ These five species may be distinguished as follows :—

* Base of stem and lower leaves spotted or streaked with brown, guiding-plates of lip forming two distinct ridges.

Terminal lobe of lip large, broadly cordate . . *Serapias cordigera*, L.
Terminal lobe of lip very small, ovato-lanceolate . *S. occultata*, Gay.

** Stem and leaves quite free from spots or streaks, guiding-plates forming two distinct ridges.

Terminal lobe of lip large, broadly cordate, lateral
lobes protruding . . . . *S. neglecta*, Dntrs.
Terminal lobe of lip lanceolate or ovato-lanceolate,
lateral lobes included within sepals or nearly so *S. longipetala*, Poll.

*** Stem and leaves quite free from spots or streaks, guiding-plates united.

Terminal lobe of lip ovate acute, oblongo-ovate,
or oblongo-lanceolate . . . . *S. lingua*, L.

The rare hybrid known as *S. triloba*, Viv., has been found, as stated in the note, at Berre (Alpes Maritimes), at Diano (between Alassio and Oneglia), and Professor de Notaris says§ that it has been found here and there on the grassy slopes near Genoa.

M. Barla, in his valuable illustrated work on the Orchids of Nice and the Department, gives some instances of other hybrids in this family—viz., *Orchis ustulato-tridentata*, Canut ; *O. tridentata militaris*, Canut ; *O. coriophoro-palustris*, Timb. Lag.

EXPLANATION OF PLATE XCIV.—Fig. A 1, labellum, of the natural size. A 2, petal, of the natural size. B 1, labellum, of the natural size. B 2, petal, of the natural size. B 3, column and part of the labellum, magnified.

† Fl. Ital. iii. 431.

‡ *Serapias papilionaceo-lingua*, Barla. "Cette plante a été trouvé près de Berre en Avril, 1866, dans une localité où croissent en assez grande abondance l'*Orchis papilionacea*, le *Serapias lingua*, et le *S. longipetala*."—Barla, ' Iconog. des Orchidées des Alpes Maritimes,' p. 34, tab. 22.

§ Repertorium, Fl. Ligusticæ, in Memorie del Acad. Sc. di Torino (1848), p. 425.

A.   B.   C.   A.1.   C.1.   C.2.   C.3.

PLATE XCV.

(A, B.) SERAPIAS LINGUA, L.; (C.) S. OCCULTATA, Gay.

Natural Order ORCHIDACEÆ.

GEN. CHAR.—See description of Plate XVI. Part 1.

(A, B.) SPEC. CHAR.—*Petals* subulate above, enlarged below into a broadly ovate base. *Lip* pinkish flesh colour, bright pink, or yellowish-brown, pubescent or subglabrous, terminal lobe ovate-acute (as in A.), or oblongo-ovate, or oblongo-lanceolate (as in B.); *guiding-plates* (calli) united, forming an oblongo-obovate fleshy mass, channelled in its posterior third only, where it becomes confluent with the walls of the stigmatic cavity. *Pollen-masses* yellow, or yellowish-green. *Leaves* and *stem* free from spots. *Tubers* usually 3, one or two of which are usually long-stalked and one sessile.

Serapias lingua, L. Sp. Plant. p. 1344; Gren. et Godr. Fl. de Fr. iii. 280; Parl. Fl. Ital. iii. 422; Ardoino, Fl. Alp. Mar. p. 358; *S. oxyglottis*, Woods, Tour. Fl. p. 355.

(C.) SPEC. CHAR.—*Flowers* very small and exceedingly fertile, nearly upright. *Lip* pinkish-brown, very small, equalling or only slightly exceeding the sepals; terminal lobe ovato-lanceolate, reflexed, pubescent; *guiding-plates* forming two distinct, parallel ridges. *Pollen-masses* nearly white, very loose in texture, apparently allowing the grains to fall. *Leaves* and *stem* spotted in the lower part of the plant, stem straight. *Tubers* 2, one sessile, one shortly stalked.

Serapias occultata, Gay, Ann. Sc. Nat., sér. 2, vol. vi. p. 119; Gren. et Godr. Fl. de Fr. iii. 280; *S. parviflora*, Parl. in Giorn. Sc. per la Sicilia (anno 1837) lix. 66, et in *Linnæa*, xii. 347, tab. iv.; Woods, Tour. Fl. p. 355; *S. laxiflora*, Chaub. var. *parviflora*, Rchb. Ic. Fl. Germ. xiii. 13, tab. 442; *S. strictiflora*, Welw. MSS. et in Herbar. Kew.

HABITATS.—(A, B.) gathered by me on Montegrosso, Mentone, April 20, 1870; (C.) gathered by my father at Ceinturon on the shore near Hyères on the 14th of May, 1867.

REMARKS.—The small flowered forms of *Serapias lingua*, L., may at first sight be readily mistaken for *S. occultata*, Gay, but I have never seen specimens in which this likeness was more than superficial. The erect habit of the stem and flowers, the specks of brownish colour at the bases of the lower leaves and the stems, the loose texture of the pollen-masses, and perhaps the short-stalked tuber, distinguish *S. occultata*. The figure of this latter plant, from specimens gathered at Mondello, given by Professor Parlatore in the *Linnæa* (vid. sup.) corresponds well with the Hyères plant. In Professor Reichenbach's figure, however, the tubers are represented as long-stalked, and the aspect of the plant would appear to be somewhat different. I am aware that the characters drawn from the tubers are said by some authorities not to be reliable. Dr. Welwitsch has appended a note to some specimens in the Kew Herbarium of *S. cordigera* (which is usually described as producing only two tubers), in which he states that the greater number of individuals of this species in a certain part of Portugal habitually produce three tubers, and more rarely only two. Here I may observe, however, that it is possible that in certain districts races may be found characterized by peculiarities in their tubers, so that we might distinguish the *Serapias cordigera*, L., as it appears in some parts of Portugal, from that of Southern France, where I have taken up large numbers of the roots of this species without finding a single example with more than two tubers. I am indebted to Dr. Welwitsch for some interesting notes on the *Serapias occultata*, Gay, which he recognises as identical with his *Serapias strictiflora*, Welw., specimens of which may be found in the Kew Herbarium. Dr. Welwitsch says, " I met with this plant first in the month of May, 1840, and more frequently in the month of May, 1841, and nearly always *in pratis brevius herbidis humidiusculis juxta rivulos, nunc in sylvis lucidis nunc in ericetis apricis prope* Lumiar *et* Bellas *in agro Olisiponensi nec non ad pedem de* ' Serra d'Arrabida' *trans Tagum.* At that time I was preparing my *Flora lusitanica exsiccata* for the *Unio itineraria Würtemburgensis*, and shortly after the distribution of my Portuguese plants by the said Union, Professor Hochstetter informed me that my *S. strictiflora* had been described by J. Gay under the name of *S. occultata*. In the following years, from 1843 to 1850, when extending my botanical excursions to the other districts of Estremadura and the neighbouring provinces of Portugal, I met with the same *Serapias* nearly everywhere in similar localities, but also sometimes on dry and sometimes on nearly *inundated* spots, and so great was the number of varieties that I became very doubtful about the real specific difference of the said *Serapias* from *S. lingua*. By *this circumstance* I have been induced to *retain* my first manuscript name for the primitive form in nearly all my latter distribution of Portuguese plants. . . . I have never published a description of my plant. . . ." We have here then high authority in favour of the doubts as to whether *Serapias occultata* is really a distinct species. In all the specimens of *Serapias occultata* which I have examined, the flowers were capable of self-fertilization, on account of the loose structure of the pollen-masses, packets from which fall spontaneously on to the stigmatic surface, producing certain impregnation, as the quantities of regularly set capsules witness. This is quite a new feature for the genus, and one which, if it proves constant on extended investigation, will afford a specific character of great physiological importance. The condition of the pollen-masses seems therefore, to be quite analogous to that in *Orchis intacta*, Lk.,[*] these species forming two of the rare exceptions in which an orchid is found to be capable of self-fertilization. *Serapias occultata*, Gay,

[*] See Darwin, ' Notes on Fert. of Orchids' in Ann. Nat. Hist. for Sept. 1869, p. 3.

grows near Gijon in the Asturias (Durieu), in Portugal (Welwitsch and others), Algiers, Tuscany, Sicily, Capri, Malta, near Athens, and on Mount Pentelicus, at Smyrna, and in the island of Cyprus (Parlatore, l. c.). M. Barla, in his lately published 'Iconographie des Orchidées des Alpes Maritimes,' gives a figure of a plant which he names *S. occultata*, Gay, and which has been found in very small quantity between Ventimiglia and Bordighelera, and on the Croisette, near Cannes, but these drawings appear to me to represent small-flowered forms of *S. lingua* and not *S. occultata*, Gay. I do not know of any station for *S. occultata*, Gay, in the region between Marseilles and Genoa other than those in the neighbourhood of Toulon and Hyères.

Serapias lingua, L., has the same east and west extension as *S. occultata*, Gay, but is far more abundant, and is the species most frequently met with along the whole Mediterranean region; it has also a greater range on the north-west, finding, with *S. cordigera*, L., its northern limit in the neighbourhood of Nantes, in lat. 47°. I have caught the same insect (Ceratina albilabris), mentioned above (see Plate XVI. Part I.), as fertilizing *Serapias cordigera*, within the hood of *S. lingua*, and also in the flowers of *Aceras anthropophora*, R. Br., and bearing their respective pollinia. A very singular abnormal development may not unfrequently be observed at Mentone in *Serapias lingua*, and *S. cordigera*, in which the anterior half of the lateral sepals has taken on the structure of the labellum.[*]

EXPLANATION OF PLATE XCV.—Fig. A 1, column and part of the labellum, showing the united guiding-plates, magnified.   B, a flower of *S. lingua*, L., from another plant, of the natural size.   C 1, column and part of the labellum, magnified.   C 2, labellum, magnified.   C 3, column viewed from in front, showing the grains of pollen falling from the anther-cells.

[*] See Journal of Linnean Soc., London (Botany), vol. xi. p. 490.

Plate XCVI.

ORCHIS papilionacea, L.

Natural Order Orchidaceae.

Gen. Char.—See description of Plate XVII. Part I.

Spec. Char.—*Spike* 2-6 flowered, lax. *Sepals* connivent below, spreading, above. *Petals* shorter, connivent, and forming a hood over the anther. *Lip* rhomboid, fanshaped, or suborbicular, often contracted and faintly lobed at the sides, crenately toothed, beautifully veined, having two prominent guiding plates (fig. 3) at the point of union with the walls of the stigmatic cavity; *spur* nearly straight, parallel to and nearly equalling or shorter by one-third than the ovary. *Anther* beaked, ascending; *glands of pollen-masses* distinct, linear oblong, each having a hood-shaped thickening at its anterior end (that nearest to an insect advancing towards the spur), about three times as long as broad. *Leaves* linear lanceolate, more or less channelled. *Tubers* 2.

Orchis papilionacea, Linn. Sp. Plant. p. 1331 ; Gren. et Godr. Fl. de Fr. iii. 284 ; Parl. Fl. Ital. iii. 458 ; Woods, Tour. Fl. p. 350 ; Ardoino, Fl. Alp. Mar. p. 351.

Habitat.—Below Castelaras near Cannes, where I gathered the specimen represented as the central one of the group on the 9th of May, 1866, and where Mr. Orr collected the two remaining specimens on May 12th, 1870, when the very dry season had dwarfed the growth of many plants.

Remarks.—It is difficult to look at a living plant of this truly beautiful *Orchis* and not be struck by the points of resemblance which it appears to present in common with the genus *Serapias*, and yet this resemblance is, for the most part, only superficial. However, there seems no doubt but that this species has asserted its affinity by crossing with certain species of *Serapias*, and that the so-called *Serapias triloba*, Viv., is the result of this union. But we must remember that hybrid plants which are believed to have had *Orchis laxiflora*, Lamck., *Serapias cordigera*, L., in one instance, and *S. longipetala*, Poll., in another, for parents, have also been recorded.* M. Barla,† describes and figures a hybrid orchid, the parents of which are believed to be *Orchis papilionacea*, L., on the one side, and *O. Morio*, L., on the other, specimens of which were found in the Contes and Bendejeun valleys, in company with the supposed parents. Unfortunately, experimental evidence of the parentage of these plants is

* Gren. et Godr. Fl. de Fr. iii. 277.
† 'Iconographie des Orchidées de Nice,' p. 44, tab. 20.

wanting, and as it has hitherto been found almost impossible to raise orchids with certainty by seed, this is rendered very difficult of attainment.

I am not aware that any description has been given of the peculiar structure of the glands, or of the character of the guiding plates, which here serve as barriers to prevent the straying of insects to the right or left of the orifice of the spur, and from the line which brings them in contact with the glands of the pollen masses, and the stigmatic cavity, which is to receive the pollen thus removed. The contraction of these curious strap-shaped glands causes a more rapid depression of the stalk of the pollen-mass than I have ever seen in any other *Orchis* or *Ophrys*. A somewhat similar oblong gland is found in *Nigritella globosa*, Rchb.

The habitats which I find recorded between Marseilles and Genoa are the following :—Esterel, Chateauneuf, Biot, Castelaras, Grasse (Rchb.), Berre, Contes and Nice in the Department Alpes Maritimes, and San Remo, Vado (Parlatore), Varazze, Sestri di Poneute, and the neighbourhood of Genoa,* where it is said to abound.

The general distribution of the species is thus given by Nyman :†— France (Lyons and Grasse) ; Portugal ; Spain, Granada ; Italy (Naples, Sicily); Carniola ; Dalmatia, Hungary, Transylvania, Greece, Turkey. To these stations may be added Toulouse (Grenier and Godron), the extreme point reached in a north-westerly direction, and Algeria (Parlatore) on its southern limit.

M. Boissier‡ states his opinion that the plant of the Mediterranean region is the *Orchis rubra* of Jacquin, which he considers distinct from *Orchis papilionacea*, L., on account of its smaller and cuneiform labellum, which grows in Portugal, Southern Spain, and Algeria. I have not been able to appreciate the distinction, though favoured by M. Boissier with a specimen of *O. papilionacea* from Constantine, and I find that M. Parlatore treats *O. rubra*, Jacq., as a mere variety of *papilionacea*.

EXPLANATION OF PLATE XCVI.—Fig. 1, lip of the rhomboid type. Fig. 2, lip, from the flower of a distinct plant, of the suborbicular type. Fig. 3, column, lower part of the lip, spur, and ovary. Fig. 4, pollen-mass with its gland adhering to the point of a pencil, which was made to play the part of an insect. Fig. 5, lower part of the caudicle (stalk of the pollen-mass) and the gland. Figs. 1 and 2 are of the natural size, and 3, 4, 5, magnified.

* De Notaris, 'Repert. Fl. Ligust.' in Acad. Sc. Torino (1846 et 1848), p. 421.
† 'Sylloge,' Fl. Eur. p. 356.   ‡ 'Voy. Bot. dans l'Espagne,' ii. 750.

Plate 87

PLATE XCVII.

CYPERUS MELANORHIZUS, Delile.

Natural Order CYPERACEÆ.

GEN. CHAR.—" Perennial, rarely annual, rushy or grass-like herbs of various habit. *Spikelets* linear, compressed, in lateral or terminal usually bracteate heads, or branched umbels or panicles. *Glumes* many, distichous, concave, keeled, deciduous, all or most flowering. *Flowers* 2-sexual. *Bristles* 0. *Stamens* 1-3. *Style* deciduous, not tumid at the base, stigmas 2-3. *Fruit* 3-gonous or compressed."—HOOKER, *Student's Flora of Brit. Is.*, p. 405.

SPEC. CHAR.—*Spikelets* arranged in a simple, or, more rarely, compound umbel, of many rays; *general involucral bracts* leafy, one or more exceeding the rays of the umbel. *Spikelets* linear lanceolate, laterally compressed; *glumes* ovate, obtuse with a short mucro, many-nerved. *Achenes* . . . . [" elliptic, triquetrous, obtuse, apiculate, punctulate under the lens (olivaceous), shorter than the glume by one-half."—*Parlatore.*]* *Stem* triquetrous. *Rhizome* stoloniferous, each stolon usually ending in an edible tuber about the size of a nut, which when young is covered with brownish scales, and short white root-like processes, but finally becomes nearly black, and is surrounded with narrow zones formed by the ridges which mark the bases of the fallen scales.

Cyperus melanorhizus, Delile, Illust. Fl. Ægyp. No. 40; *Cyp. aureus*, Ten. Fl. Nap. Prodr. p. 8; Gren. et Godr. Fl. de Fr. iii. 360; Ardoino, Fl. Alp. Mar. p. 394; *Cyp. Tenorii*, Presl. Fl. Sic. p. 43; Woods, Tour. Fl. p. 360 (part).

REMARKS.—Professor Parlatore, and other botanists who have had opportunities of examining *Cyperus esculentus*, Linn., in cultivation, are of opinion that it should probably be treated as a domesticated variety of *Cyp. melanorhizus*, Del. This latter plant grows at Mentone as a weed of terraced ground devoted to lemons and oranges; and appears to be there dependent on its tubers for multiplication, for I have never been able to find a single ripe seed.

Still I cannot learn that it has ever been cultivated by the peasants for the sake of its nut-flavoured tubers,† as the esculent variety is in southern Italy, Spain, and Syria. At Pegli, near Genoa, *Cyp. melanorhizus* is stated to grow in the garden of the Villa Grimaldi, and there is a specimen at Kew from Mabille's Herbarium Corsicum, No. 235, gathered

---

* Parlatore, Fl. Ital. ii. 33.

† The tubers of the Mentonese plant are quite sweet and nutty, and not as in the Corsican and "Toulon" plant, described by MM. Grenier and Godron, bitter. The supposed existence of this species at Toulon has, I believe, not been confirmed.

in Corsica, in the "cultures d'Erba longa," Oct. 1867. Professor Parlatore, however, says that in southern Italy this plant is found near the sea in "luoghi erbosi o arenose ;"* implying, I imagine, that it grows in uncultivated ground.

According to the same authority, *Cyp. melanorhizus* is found at Mentone, Pegli, Ostia, Naples, Sicily, and the adjacent islands, and in Egypt. To this we may add, judging from specimens in the Kew herbarium, Bagdad (collected by Noë, and communicated by H. G. Reichenbach), Crete (Sieber), and Lenkoran, in Persia, on the south-western shore of the Caspian Sea (Dr. Fischer).

Two plants of great interest and rarity, belonging to the Cyperus (sedge) family have lately been discovered by my father and myself, in the marshy ground about the mouth of the Roya, at Ventimiglia. These are *Cyperus globosus*, All., and *Fimbrystilis annua*, R. et S. Of the former, hitherto only known to grow near Nice, Verona, in southern Spain, Arabia, and eastern India, we have repeatedly gathered large quantities, but of the latter only a single specimen, brought in by chance along with a mass of *Cyp. globosus*, has been discovered. *F. annua* is considered a doubtful native in Europe by Parlatore, who assigns the following actual distribution for the species : tropical America (whence he thinks it may have been introduced), Switzerland, the Italian Tyrol, Venetia, Piedmont, Lombardy, Tuscany, at Smyrna, and in the island of Sara, in the Caspian Sea. When balancing the probabilities as to the introduction by human agency of this species into Europe, we must remember the fact, directly favouring the supposition, that *Euphorbia Preslii* has certainly been thus imported from North America, and is now found growing as a weed at Ventimiglia, close to the very spot where *F. annua* was discovered. On the other hand, if *F. annua* is a native of an island in the Caspian, any argument as to its being foreign to Europe, drawn from the distance between the European and American habitats, falls to the ground.

I have satisfied myself that *Cyp. globosus* may easily be distinguished from *Cyp. flavescens*, Linn., with which it grows mixed, by its ripe achenes. Those of *Cyp. flavescens* are nearly black and subglobose, and may roughly be compared with minute shot ; while those of *Cyp. globosus* are brown, subelliptic, laterally confpressed and more nearly resemble minute grains of corn convex on either side.

EXPLANATION OF PLATE XCVII.—Fig. 1, an entire spikelet, magnified. Fig. 2, side view of a glume, magnified. Fig. 3, pistil, magnified. Figs. 4, 5, and 6, the tuber in different stages of development, magnified.

* Fl. Ital. ii. 33.

INDEX.

INDEX.

LIST OF WORKS

ON

BOTANY, ENTOMOLOGY, CONCHOLOGY,

TRAVELS, TOPOGRAPHY,

ANTIQUITY, AND MISCELLANEOUS

LITERATURE AND SCIENCE.

PUBLISHED BY

L. REEVE AND CO.,

5, HENRIETTA STREET, COVENT GARDEN, W.C.

NEW SERIES OF POPULAR NATURAL
HISTORY FOR BEGINNERS AND
AMATEURS.

British Insects; a Familiar Description of the Form, Structure, Habits, and Transformations of Insects. By E. F. STAVELEY. Crown 8vo, 16 Coloured Plates, and numerous Wood Engravings, 14s.

British Butterflies and Moths; an Introduction to the Study of our Native LEPIDOPTERA. By H. T. STAINTON. Crown 8vo, 16 Coloured Plates, and Wood Engravings, 10s. 6d.

British Beetles; an Introduction to the Study of our indigenous COLEOPTERA. By E. C. RYE. Crown 8vo, 16 Coloured Plates, and 11 Wood Engravings, 10s. 6d.

British Bees; an Introduction to the Study of the Natural History and Economy of the Bees indigenous to the British Isles. By W. E. SHUCKARD. Crown 8vo, 16 Coloured Plates, and Woodcuts, 10s. 6d.

British Spiders; an Introduction to the Study of the ARANEIDÆ found in Great Britain and Ireland. By E. F. STAVELEY. Crown 8vo, 16 Coloured Plates, and 44 Wood Engravings, 10s. 6d.

British Grasses; an Introduction to the Study of the Grasses found in the British Isles. By M. PLUES. Crown 8vo, 16 Coloured Plates, and 100 Wood Engravings, 10s. 6d.

British Ferns; an Introduction to the Study of the FERNS, LYCOPODS, and EQUISETA indigenous to the British Isles. With Chapters on the Structure, Propagation, Cultivation, Diseases, Uses, Preservation, and Distribution of Ferns. By M. PLUES. Crown 8vo, 16 Coloured Plates, and 55 Wood Engravings, 10s. 6d.

British Seaweeds; an Introduction to the Study of the Marine ALGÆ of Great Britain, Ireland, and the Channel Islands. By S. O. GRAY. Crown 8vo, 16 Coloured Plates, 10s. 6d.

BOTANY.

The Natural History of Plants. By H. BAILLON, President of the Linnæan Society of Paris, Professor of Medical Natural History and Director of the Botanical Garden of the Faculty of Medicine of Paris. Super-royal 8vo. Vols I. to V., with 2300 Wood Engravings, 25s. each.

Handbook of the British Flora; a Description of the Flowering Plants and Ferns indigenous to, or naturalized in, the British Isles. For the use of Beginners and Amateurs. By GEORGE BENTHAM, F.R.S. 4th Edition, revised, Crown 8vo, 12s.

Illustrations of the British Flora; a Series of Wood Engravings, with Dissections, of British Plants, from Drawings by W. H. FITCH, F.L.S., and W. G. SMITH, F.L.S., forming an Illustrated Companion to BENTHAM's "Handbook," and other British Floras. 1306 Wood Engravings, 12s.

Domestic Botany; an Exposition of the Structure and Classification of Plants, and of their uses for Food, Clothing, Medicine, and Manufacturing Purposes. By JOHN SMITH, A.L.S., ex-Curator of the Royal Gardens, Kew. Crown 8vo, 16 Coloured Plates and Wood Engravings, 16s.

British Wild Flowers, Familiarly Described in the Four Seasons. By THOMAS MOORE, F.L.S. 24 Coloured Plates, 16s.

The Narcissus, its History and Culture, with Coloured Figures of all known Species and Principal Varieties. By F. W. BURBIDGE, and a Review of the Classification by J. G. BAKER, F.L.S. Super-royal 8vo, 48 Coloured Plates, 32s.

The Botanical Magazine; Figures and Descriptions
of New and Rare Plants of interest to the Botanical Student, and
suitable for the Garden, Stove, or Greenhouse. By Sir J. D.
HOOKER, K.C.S.I., C.B., F.R.S., Director of the Royal Gardens,
Kew. Royal 8vo. Third Series, Vols. I. to XXXV., each 42s.
Published Monthly, with 6 Plates, 3s. 6d., coloured. Annual
Subscription, 42s.

RE-ISSUE of the THIRD SERIES in Monthly Vols., 42s. each; to
Subscribers for the entire Series, 36s. each.

The Floral Magazine; New Series, Enlarged to
Royal 4to. Figures and Descriptions of the choicest New Flowers
for the Garden, Stove, or Conservatory. Vols. I. to VIII., in handsome cloth, gilt edges, 42s. each. Monthly, with 4 beautifully-
coloured Plates, 3s. 6d. Annual Subscription, 42s.

FIRST SERIES complete in Ten Vols., with 560 beautifully-coloured
Plates. £18 7s. 6d.

The Young Collector's Handybook of Botany.
By the Rev. H. P. DUNSTER, M.A. 66 Wood Engravings,
3s. 6d.

Laws of Botanical Nomenclature adopted by
the International Botanical Congress, with an Historical Introduction and a Commentary. By ALPHONSE DE CANDOLLE.
2s. 6d.

Contributions to the Flora of Mentone, and to a
Winter Flora of the Riviera, including the Coast from Marseilles
to Genoa. By J. TRAHERNE MOGGRIDGE, F.L.S. Royal 8vo.
Complete in One Vol., with 99 Coloured Plates, 63s.

Flora Vitiensis; a Description of the Plants of
the Viti or Fiji Islands, with an Account of their History, Uses,
and Properties. By Dr. BERTHOLD SEEMANN, F.L.S. Royal
4to, Coloured Plates, £8 5s.

Flora of Mauritius and the Seychelles; a Descrip-
tion of the Flowering Plants and Ferns of those Islands. By
J. G. BAKER, F.L.S. 24s. Published under the authority of the
Colonial Government of Mauritius.

Flora of British India. By Sir. J. D. Hooker,
K.C.S.I., C.B., F.R.S., &c.; assisted by various Botanists. Parts
I. to VI., 10s. 6d. each. Vols. I. & II., cloth, 32s. each.
Published under the authority of the Secretary of State for India
in Council.

Flora of Tropical Africa. By Daniel Oliver,
F.R.S., F.L.S. Vols. I. to III., 20s. each. Published under the
authority of the First Commissioner of Her Majesty's Works.

Handbook of the New Zealand Flora; a Systematic
Description of the Native Plants of New Zealand, and the
Chatham, Kermadec's, Lord Auckland's, Campbell's, and Mac-
quarrie's Islands. By Sir J. D. Hooker, K.C.S.I., F.R.S.
Complete in One Vol., 30s. Published under the auspices of the
Government of that Colony.

Flora Australiensis; a Description of the Plants
of the Australian Territory. By George Bentham, F.R.S.,
assisted by Ferdinand Mueller, F.R.S., Government Botanist,
Melbourne, Victoria. Complete in Seven Vols., £7 4s. Vols. I.
to VI., 20s. each; Vol. VII., 24s. Published under the auspices
of the several Governments of Australia.

Flora of the British West Indian Islands. By
Dr. Grisebach, F.L.S. 37s. 6d. Published under the auspices
of the Secretary of State for the Colonies.

Flora Hongkongensis; a Description of the
Flowering Plants and Ferns of the Island of Hongkong. By
George Bentham, F.R.S. With a Map of the Island, and a
Supplement by Dr. Hance. 18s. Published under the authority
of Her Majesty's Secretary of State for the Colonies. The Sup-
plement separately, 2s. 6d.

Flora Capensis; a Systematic Description of the
Plants of the Cape Colony, Caffraria, and Port Natal. By
William H. Harvey, M.D., F.R.S., Professor of Botany in
the University of Dublin, and Otto Wilhem Sonder, Ph.D.
Vols. I. and II., 12s. each. Vol. III., 18s.

Elementary Lessons in Botanical Geography. By
J. G. Baker, F.L.S. 3s.

On the Flora of Australia: its Origin, Affinities,
and Distribution; being an Introductory Essay to the "Flora of
Tasmania." By Sir J. D. Hooker, F.R.S. 10s.

Genera Plantarum, ad Exemplaria imprimis in
Herbariis Kewensibus servata definita. By George Bentham,
F.R.S., F.L.S., and Sir J. D. Hooker, F.R.S., Director of the
Royal Gardens, Kew. Vol. I.—Part I., Royal 8vo, 21s.; Part
II., 14s.; Part III., 15s.; or Vol. I. complete, 50s. Vol. II.—
Part I., 24s.; Part II. 32s.; or Vol. II. complete, 56s.

Illustrations of the Genus Carex. By Francis
Boott, M.D. Folio, 600 Plates. Part I., £12; Parts II. and
III., £6 each; Part IV., £12.

Illustrations of the Nueva Quinologia of Pavon,
with Observations on the Barks described. By J. E. Howard,
F.L.S. With 27 Coloured Plates. Imperial folio, half-morocco.
gilt edges, £6 6s.

The Quinology of the East Indian Plantations.
By J. E. Howard, F.L.S. Complete in One Vol., folio. With 13
Coloured and 2 Plain Plates, and 2 Photo-prints, 84s. Parts
II. and III., cloth, 63s.

Revision of the Natural Order Hederaceæ; being
a reprint, with numerous additions and corrections, of a series of
papers published in the "Journal of Botany, British and Foreign."
By Berthold Seemann, Ph.D., F.L.S. 7 Plates, 10s. 6d.

Icones Plantarum. Figures, with Brief Descriptive Characters and Remarks, of New and Rare Plants, selected
from the Author's Herbarium. By Sir W. J. Hooker, F.R.S.
New Series, Vol. V. 100 Plates, 31s. 6d.

Botanical Names for English Readers. By Randal
H. Alcock. 8vo, 6s.

Orchids; and How to Grow them in India and
other Tropical Climates. By SAMUEL JENNINGS, F.L.S., F.R.H.S,.
late Vice-President of the Agri-Horticultural Society of India.
Royal 4to. Complete in One Vol., cloth, gilt edges, 63s.

A Second Century of Orchidaceous Plants, selected
from the Subjects published in Curtis's "Botanical Magazine"
since the issue of the "First Century." Edited by JAMES BATE-
MAN, Esq., F.R.S. Complete in One Vol., Royal 4to, 100 Coloured
Plates, £5 5s.

Dedicated by Special Permission to H.R.H. the Princess of Wales.

Monograph of Odontoglossum, a Genus of the
Vandeous Section of Orchidaceous Plants. By JAMES BATEMAN,
Esq., F.R.S. Imperial folio, complete in Six Parts, each with 5
Coloured Plates, and occasional Wood Engravings, 21s.; or, in
One Vol., half-morocco, gilt edges, £7 7s.

Select Orchidaceous Plants. By ROBERT WARNER,
F.R.H.S. With Notes on Culture by B. S. WILLIAMS. Folio,
with 40 Coloured Plates, cloth gilt. £7 7s.
Second Series, complete, with 39 Coloured Plates, £7 7s.
Third Series, Parts I. to III., 10s. 6d. each.

The Rhododendrons of Sikkim-Himalaya; being
an Account, Botanical and Geographical, of the Rhododendrons
recently discovered in the Mountains of Eastern Himalaya, by
Sir J. D. Hooker, F.R.S. By Sir W. J. HOOKER, F.R.S. Folio,
30 Coloured Plates, £4 14s. 6d.

Outlines of Elementary Botany, as Introductory
to Local Floras. By GEORGE BENTHAM, F.R.S., President of
the Linnæan Society. New Edition, 1s.

British Grasses; an Introduction to the Study
of the Gramineæ of Great Britain and Ireland. By M. PLUES.
Crown 8vo, with 16 Coloured Plates and 100 Wood Engravings,
10s. 6d.

FERNS.

British Ferns; an Introduction to the Study of the FERNS. LYCOPODS, and EQUISETA indigenous to the British Isles. With Chapters on the Structure, Propagation, Cultivation, Diseases, Uses, Preservation, and Distribution of Ferns. By M. PLUES. Crown 8vo, with 16 Coloured Plates, and 55 Wood Engravings, 10s. 6d.

The British Ferns; Coloured Figures and Descriptions, with Analysis of the Fructification and Venation of the Ferns of Great Britain and Ireland. By Sir W. J. HOOKER, F.R.S. Royal 8vo, 66 Coloured Plates, £2 2s.

Garden Ferns; Coloured Figures and Descriptions with Analysis of the Fructification and Venation of a Selection of Exotic Ferns, adapted for Cultivation in the Garden, Hothouse, and Conservatory. By Sir W. J. HOOKER, F.R.S. Royal 8vo, 64 Coloured Plates, £2 2s.

Filices Exoticæ; Coloured Figures and Description of Exotic Ferns. By Sir W. J. HOOKER, F.R.S. Royal 4to, 100 Coloured Plates, £6 11s.

Ferny Combes; a Ramble after Ferns in the Glens and Valleys of Devonshire. By CHARLOTTE CHANTER. Third Edition. Fcap. 8vo, 8 Coloured Plates and a Map of the County, 5s.

MOSSES.

Handbook of British Mosses, containing all that are known to be natives of the British Isles. By the Rev. M. J. BERKELEY, M.A., F.L.S. 24 Coloured Plates, 21s.

Synopsis of British Mosses, containing Descrip-tions of all the Genera and Species (with localities of the rarer ones) found in Great Britain and Ireland. By CHARLES P. HOBKIRK, President of the Huddersfield Naturalists' Society. Crown 8vo, 7s. 6d.

SEAWEEDS.

British Seaweeds; an Introduction to the Study of
the Marine ALGÆ of Great Britain, Ireland, and the Channel
Islands. By S. O. GRAY. Crown 8vo, with 16 Coloured Plates,
10s. 6d.

Phycologia Britannica; or, History of British
Seaweeds. Containing Coloured Figures, Generic and Specific
Characters, Synonyms and Descriptions of all the Species of Algæ
inhabiting the Shores of the British Islands. By Dr. W. H.
HARVEY, F.R.S. New Edition. Royal 8vo, 4 vols. 360
Coloured Plates, £7 10s.

Phycologia Australica; a History of Australian
Seaweeds, comprising Coloured Figures and Descriptions of the
more characteristic Marine Algæ of New South Wales, Victoria,
Tasmania, South Australia, and Western Australia, and a
Synopsis of all known Australian Algæ. By Dr. W. H. HARVEY,
F.R.S. Royal 8vo, Five Vols., 300 Coloured Plates, £7 13s.

FUNGI.

Outlines of British Fungology, containing Cha-
racters of above a Thousand Species of Fungi, and a Complete
List of all that have been described as Natives of the British Isles.
By the Rev. M. J. BERKELEY, M.A., F.L.S. 24 Coloured Plates,
30s.

The Esculent Funguses of England. Containing
an Account of their Classical History, Uses, Characters, Development, Structure, Nutritious Properties, Modes of Cooking and
Preserving, &c. By C. D. BADHAM, M.D. Second Edition.
Edited by F. CURREY, F.R.S. 12 Coloured Plates, 12s.

Illustrations of British Mycology, comprising
Figures and Descriptions of the Funguses of interest and novelty
indigenous to Britain. By Mrs. T. J. HUSSEY. Royal 4to,
Second Series, 50 Coloured Plates, £4 10s.

Clavis Agaricinorum; an Analytical Key to the
British Agaricini, with Characters of the Genera and Sub-genera.
By WORTHINGTON G. SMITH, F.L.S.   6 Plates, 2s. 6d.

SHELLS AND MOLLUSKS.

Testacea Atlantica; or, the Land and Freshwater
Shells of the Azores, Madeiras, Salvages, Canaries, Cape Verdes,
and Saint Helena. By T. VERNON WOLLASTON, M.A., F.L.S.
Demy 8vo, 25s.

Elements of Conchology; an Introduction to
the Natural History of Shells, and of the Animals which form
them. By LOVELL REEVE, F.L.S. Royal 8vo, Two Vols., 62
Coloured Plates, £2 16s.

Conchologia Iconica; or Figures and Descriptions
of the Shells of Mollusks, with remarks on their Affinities, Synonymy, and Geographical Distribution. By LOVELL REEVE,
F.L.S., and G. B. SOWERBY, F.L.S., complete in Twenty Vols.,
4to, with 2727 Coloured Plates, half-calf, £178.

A detailed list of Monographs and Volumes may be had.

Conchologia Indica; Illustrations of the Land and
Freshwater Shells of British India. Edited by SYLVANUS
HANLEY, F.L.S., and WILLIAM THEOBALD, of the Geological
Survey of India. Complete in One Vol., 4to, with 160 Coloured
Plates, £8 5s.

The Edible Mollusks of Great Britain and Ireland,
with the Modes of Cooking them. By M. S. LOVELL. Crown
8vo, with 12 Coloured Plates, 8s. 6d.

INSECTS.

Insecta Britannica; Vol. III., Diptera. By
FRANCIS WALKER, F.L.S. 8vo, with 10 Plates, 25s.

The Larvæ of the British Lepidoptera, and their
Food Plants. By OWEN S. WILSON. With Life-size Figures,
drawn and coloured from Nature by ELEANORA WILSON. Super-
royal 8vo. With 40 elaborately-coloured Plates, 63s.

British Insects. A Familiar Description of the
Form, Structure, Habits, and Transformations of Insects. By
E. F. STAVELEY, Author of "British Spiders." Crown 8vo, with
16 Coloured Plates and numerous Wood Engravings, 14s.

British Beetles ; an Introduction to the Study
of our indigenous COLEOPTERA. By E. C. RYE. Crown 8vo,
16 Coloured Steel Plates, and 11 Wood Engravings, 10s. 6d.

British Bees; an Introduction to the Study of the
Natural History and Economy of the Bees indigenous to the
British Isles. By W. E. SHUCKARD. Crown 8vo, 16 Coloured
Plates, and Woodcuts of Dissections, 10s. 6d.

British Butterflies and Moths ; an Introduction to
the Study of our Native LEPIDOPTERA. By H. T. STAINTON.
Crown 8vo, 16 Coloured Plates, and Wood Engravings, 10s. 6d.

British Spiders ; an Introduction to the Study of
the ARANEIDÆ found in Great Britain and Ireland. By E. F.
STAVELEY. Crown 8vo, 16 Coloured Plates, and 44 Wood
Engravings, 10s. 6d.

Harvesting Ants and Trap-door Spiders ; Notes
and Observations on their Habits and Dwellings. By J. T.
MOGGRIDGE, F.L.S. With a SUPPLEMENT of 160 pp. and 8
additional Plates, 17s. The Supplement separately, cloth,
7s. 6d.

Curtis's British Entomology. Illustrations and
Descriptions of the Genera of Insects found in Great Britain
and Ireland, Containing Coloured Figures, from Nature, of the
most rare and beautiful Species, and in many instances, upon the
plants on which they are found. Eight Vols., Royal 8vo, 770
Coloured Plates, £28.

Or in Separate Monographs

Orders.	Plates.	£	s.	d.	Orders.	Plates.	£	s.	d.
Aphaniptera ..	2	0	2	0	Hymenoptera ..	125	6	5	0
Coleoptera ...	256	12	16	0	Lepidoptera ..	193	9	13	0
Dermaptera...	1	0	1	0	Neuroptera ...	13	0	13	0
Dictyoptera...	1	0	1	0	Omaloptera...	6	0	6	0
Diptera	103	5	3	0	Orthoptera...	5	0	5	0
Hemiptera ...	32	1	12	0	Strepsiptera ..	3	0	3	0
Homoptera ..	21	1	1	0	Trichoptera ..	9	0	9	0

"Curtis's Entomology," which Cuvier pronounced to have "reached the ultimatum of perfection," is still the standard work on the Genera of British Insects. The Figures executed by the author himself, with wonderful minuteness and accuracy, have never been surpassed, even if equalled. The price at which the work was originally published was £43 16s.

ANTIQUARIAN.

Sacred Archæology; a Popular Dictionary of Ecclesiastical Art and Institutions from Primitive to Modern Times. Comprising Architecture, Music, Vestments, Furniture Arrangement, Offices, Customs, Ritual Symbolism, Ceremonial Traditions, Religious Orders, &c., of the Church Catholic in all ages. By MACKENZIE E. C. WALCOTT, B.D. Oxon., F.S.A., Precentor and Prebendary of Chichester Cathedral. Demy 8vo, 18s.

A Manual of British Archæology. By CHARLES BOUTELL, M.A. 20 Coloured Plates, 10s. 6d.

The Antiquity of Man; an Examination of Sir Charles Lyell's recent Work. By S. R. PATTISON, F.G.S. Second Edition. 8vo, 1s.

MISCELLANEOUS.

Report on the Forest Resources of Western Australia. By Baron FERD. MUELLER, C.M.G., M.D., Ph.D., F.R.S., Government Botanist of Victoria. Royal 4to, 20 Plates, 12s.

West Yorkshire; an Account of its Geology, Physical
Geography, Climatology, and Botany. By J. W. DAVIS, F.L.S.,
and F. ARNOLD LEES, F.L.S. Second Edition, 8vo, 21 Plates,
many Coloured, and 2 large Maps, 21s.

Handbook of the Freshwater Fishes of India;
giving the Characteristic Peculiarities of all the Species at
present known, and intended as a guide to Students and District
Officers. By Capt. R. BEAVAN, F.R.G.S., Demy 8vo, 12 plates.
10s. 6d.

Natal; a History and Description of the Colony,
including its Natural Features, Productions, Industrial Condition
and Prospects. By HENRY BROOKS, for many years a resident.
Edited by Dr. R. J. MANN, F.R.A.S., F.R.G.S., late Superin-
tendent of Education in the Colony. Demy 8vo, with Maps,
Coloured Plates, and Photographic Views, 21s.

St. Helena. A Physical, Historical, and Topo-
graphical Description of the Island, including its Geology, Fauna,
Flora, and Meteorology. By J. C. MELLISS, A.I.C.E., F.G S.,
F.L.S. In one large Vol., Super-royal 8vo, with 56 Plates and
Maps, mostly coloured, 42s.

Lahore to Yarkand. Incidents of the Route and
Natural History of the Countries traversed by the Expedition of
1870, under T. D. FORSYTH, Esq., C.B. By GEORGE HENDERSON,
M.D., F.L.S., F.R.G.S., and ALLAN O. HUME, Esq., C.B., F.Z.S.
With 32 Coloured Plates of Birds, 6 of Plants, 26 Photographic
Views, Map, and Geological Sections, 42s.

The Birds of Sherwood Forest; with Observations
on their Nesting, Habits, and Migrations. By W. J. STERLAND.
Crown 8vo, 4 plates. 7s. 6d., coloured.

The Naturalist in Norway; or, Notes on the
Wild Animals, Birds, Fishes, and Plants of that country, with
some account of the principal Salmon Rivers. By the Rev. J.
BOWDEN, LL.D. Crown 8vo, 8 Coloured Plates. 10s. 6d.

The Young Collector's Handy Book of Recreative
Science. By the Rev. H. P. DUNSTER, M.A. Cuts, 3s. 6d.

The Zoology of the Voyage of H.M.S. *Samarang*,
under the command of Captain Sir Edward Belcher, C.B., during
the Years 1843-46. By Professor OWEN, Dr. J. E. GRAY, Sir J.
RICHARDSON, A. ADAMS, L. REEVE, and A. WHITE. Edited by
ARTHUR ADAMS, F.I.S. Royal 4to, 55 Plates, mostly coloured,
£3 10s.

A Survey of the Early Geography of Western
Europe, as connected with the First Inhabitants of Britain, their
Origin, Language, Religious Rites, and Edifices. By HENRY
LAWES LONG, Esq. 8vo, 6s.

The Geologist. A Magazine of Geology, Palæont-
ology, and Mineralogy. Illustrated with highly-finished Wood
Engravings. Edited by S. J. MACKIE, F.G.S., F.S.A. Vols.
V. and VI., each with numerous Wood Engravings, 18s. Vol.
VII., 9s.

Everybody's Weather-Guide. The use of Meteoro-
logical Instruments clearly explained, with directions for secur-
ing at any time a probable Prognostic of the Weather. By A.
STEINMETZ, Esq., Author of " Sunshine and Showers," &c., 1s.

The Artificial Production of Fish. By PISCARIUS.
Third Edition. 1s.

The Gladiolus : its History, Cultivation, and Exhibition. By the Rev. H. HONYWOOD DOMBRAIN, B.A. 1s.

Zoology. By ADRIAN J. EBELL, Ph.B., M.D.
Part I., Structural Distinctions, Functions, and Classification
of the Orders of Animals, 1s.

Phosphorescence; or, the Emission of Light by
Minerals, Plants, and Animals. By Dr. T. L. PHIPSON, F.C.S.
Small 8vo, 30 Cuts and Coloured Frontispiece, 5s.

Meteors, Aerolites, and Falling Stars. By Dr. T.
L. PHIPSON, F.C.S. Crown 8vo, 25 Woodcuts and Lithographic
Frontispiece, 6s.

Papers for the People. By ONE OF THEM. No. 1,
Our Land. 8vo, 6d. (By Post, 7d. in stamps.)

The Royal Academy Album; a Series of Photo-
graphs from Works of Art in the Exhibition of the Royal Academy
of Arts, 1875. Atlas 4to, with 32 fine Photographs, cloth,
gilt edges, £6 6s.; half-morocco, £7 7s.
The same for 1876, with 48 beautiful Photo-prints, cloth,
£6 6s.; half-morocco, £7 7s. Small Edition, Royal 4to, cloth, gilt
edges, 63s.

On Intelligence. By H. TAINE, D.C.L. Oxon.
Translated from the French by T. D. HAYE, and revised, with
additions, by the Author. Complete in One Vol., 18s.

The Reasoning Power in Animals. By the Rev.
J. S. WATSON, M.A. Crown 8vo, 9s.

Manual of Chemical Analysis, Qualitative and
Quantitative; for the use of Students. By Dr. HENRY M. NOAD,
F.R.S. New Edition. Crown 8vo, 109 Wood Engravings, 16s.
Or, separately, Part I., "QUALITATIVE," New Edition, new
Notation, 6s.; Part II., "QUANTITATIVE," 10s. 6d.

Live Coals; or, Faces from the Fire. By L.
M. BUDGEN, "Acheta," Author of "Episodes of Insect Life,"
&c. Dedicated, by Special Permission, to H.R.H. Field Marshal
the Duke of Cambridge. Royal 4to, 35 Original Sketches printed
in colours, 21s.

Caliphs and Sultans; being tales omitted in the
ordinary English version of "The Arabian Nights' Entertain-
ments," freely rewritten and rearranged. By S. HANLEY, F.L.S.
6s.

PLATES.

Floral Plates, from the Floral Magazine. Beautifully Coloured, for Screens, Scrap-books, Studies in Flower-painting,
&c. 6d. and 1s. each. Lists of over 700 varieties, One Stamp.

Botanical Plates, from the Botanical Magazine.
Beautifully-coloured Figures of new and rare Plants. 6d. and 1s.
each. Lists of over 2000, One Stamp.

SERIALS.

The Botanical Magazine. Figures and Descriptions of New and rare Plants. By Sir J. D. HOOKER, C.B.,
F.R.S. Monthly, with 6 Coloured Plates, 3s. 6d. Annual
subscription, post free, 42s.

Re-issue of the Third Series, in Monthly Vols., 42s. each; to Subscribers for the entire Series, 36s. each.

The Floral Magazine. New Series, enlarged
to Royal 4to. Figures and Descriptions of Select New Flowers
for the Garden, Stove, or Conservatory. Monthly, with 4 Coloured
Plates, 3s. 6d. Annual Subscription, post free, 42s.

Select Orchidaceous Plants. By ROBERT WARNER.
Third Series. 3 Coloured Plates, 10s. 6d.

FORTHCOMING WORKS.

The Lepidoptera of Ceylon. By F. MOORE.

Genera Plantarum. By BENTHAM and HOOKER.
Vol. III., Part I.

Flora of India. By Sir J. D. HOOKER and others.
Part VII.

Natural History of Plants. By Prof. BAILLON.
Vol. VI.

Flora of Tropical Africa. By Prof. OLIVER.

Flora Capensis. By Prof. DYER.

London :

L. REEVE & CO., 5, HENRIETTA STREET, COVENT GARDEN.

GILBERT AND RIVINGTON, PRINTERS, ST. JOHN'S SQUARE, LONDON.

www.ingramcontent.com/pod-product-compliance
Lightning Source LLC
Chambersburg PA
CBHW021346210326
41599CB00011B/767